The Multi Business Model Innovation Approach

RIVER PUBLISHERS SERIES IN MULTI BUSINESS MODEL INNOVATION, TECHNOLOGIES AND SUSTAINABLE BUSINESS

Indexing: all books published in this series are submitted to the Web of Science Book Citation Index (BkCI), to CrossRef and to Google Scholar

The River Publishers Series in Multi Business Model Innovation, Technologies and Sustainable Business includes the theory and use of multi business model innovation, technologies and sustainability involving typologies, ontologies, innovation methods and tools for multi business models, and sustainable business and sustainable innovation. The series cover cross technology business modeling, cross functional business models, network based business modeling, Green Business Models, Social Business Models, Global Business Models, Multi Business Model Innovation, interdisciplinary business model innovation. Strategic Business Model Innovation, Business Model Innovation Leadership and Management, technologies and software for supporting multi business modeling, Multi business modeling and strategic multi business modeling in different physical, digital and virtual worlds and sensing business models. Furthermore the series includes sustainable business models, sustainable & social innovation, CSR & sustainability in businesses and social entrepreneurship.

Key topics of the book series include:

- Multi business models
- Network based business models
- Open and closed business models
- Multi Business Model eco systems
- Global Business Models
- Multi Business model Innovation Leadership and Management
- Multi Business Model innovation models, methods and tools
- Sensing Multi Business Models
- Sustainable business models
- Sustainability & CSR in businesses
- Sustainable & social innovation
- Social entrepreneurship and -intrapreneurship

For a list of other books in this series, www.riverpublishers.com

The Multi Business Model Innovation Approach

Editor
Peter Lindgren

Published 2018 by River Publishers
River Publishers
Alsbjergvej 10, 9260 Gistrup, Denmark
www.riverpublishers.com

Distributed exclusively by Routledge
4 Park Square, Milton Park, Abingdon, Oxon OX14 4RN
605 Third Avenue, New York, NY 10017, USA

The Multi Business Model Innovation Approach / by Peter Lindgren.

ISBN 978-87-93609-66-2 (print)

Dedication

I would like to share with you a little journey – some would call it a small BMI story – of a very dear friend of mine who is not here anymore.

I came to know him just after I returned with my family from a research stay at Polytechnico di Milan in Italy. Thanks again to Professor Marianno Corso and Professor Harry Boer for giving me that opportunity.

After a long discussion in our family my friend was named with the Italian name for wolf – Lupo.

Since my last publication and book our dear friend Lupo passed away. He was indeed a good dog, fully understood me and followed me and my family through everything and every day.

He accompanied me, Anne-Birgitte, Frederikke, Amalie and Thorbjørn from Denmark to Silicon Valley and back again. He did the Route 66 all the way from New York to Palo Alto sitting on the back seat of our car.

Lupo was always a happy and kind dog, but when cats and skunks entered his "business model ecosystem" (BMES) he became very angry, trying to protect what he thought was *his* BMES.

Lupo having just arrived at Palazzo Del Svinbovej.

Lupo at Half Moon Bay, Skyline Boulevard in Silicon Valley and driving Route 66 in the back seat of the car with Thorbjørn.

Lupo is now resting at his very favourite place at Palazzo Del Svinbovej, where he could supervise his business – Lupo's Place.

This book was written in remembrance of my good friend Lupo.

Lupo lying on his favourite place – "Lupo's Place" – at Palazzo Del Svinbovej.

Contents

Preface

I got the opportunity to write The Multi Business Model Approach (Part 1) and The Multi Business Model Innovation Approach (Part 2) at my research stay in 2016–2017 at University Tor Vergata, Italy at the CTIF Global Capsule (www.ctifglobalcapsule.com) research centre at Villa Mondracone (Picture 0.1). The stay was funded by the Aarhus University – Mobility Fund, whose support I was honoured and grateful to receive.

I have long been waiting to finish this book, which we – Post Doc Ole Horn Rasmussen and I have – began writing back in 2012. We have not had the time and research resources to finish it before now, but finally here it is after a heavy – but kindly – "push" from River Publishers and our dear friend, joint CEO Rajeev Prasad.

During the stay in Italy I got the opportunity to reflect on my team's 2005 to 2017 research work on business models and business model innovation, while participating simultaneously in other research projects and the establishment of CTIF Global Capsule (CGC).

Picture 0.1 Villa Mondracone in Frascatti, Lazio, Italy, residence of University Tor Vergata, Rome, Italy and the CTIF Global Capsule.

Pictures 0.2, 0.3, 0.4 and 0.5 The CTIF Global Capsule team behind my research stay at University Tor Vergata, Rome, Italy.

This laid the groundwork to finish this new book that you are now holding in your hand, perhaps reading in a digital version, or both. I hope you will enjoy some hours focusing on "The Multi Business Model Approach" and "The Multi Business Model Innovation Approach".

I begin the preface of this book with thanking CTIF Global Capsule President Ramjee Prasad (Picture 0.2) for making my research stay possible. I deeply thank Professor Marianna Ruggeri (Picture 0.5), Professor Massimo Colletta (Picture 0.3), Associate Professor Ernestina Cianca (Picture 0.4), the co-founders of CTIF Global Capsule and the whole research team at the University of Tor Vergata, Rome, Italy and CTIF Global Capsule (CGC 2016) for this valuable stay at your nice university. Thank you for letting me have this opportunity to "nail" down some of our work and some of my research and thoughts.

Professor
Peter Lindgren

List of Illustrations

Figures

Tables

Pictures

List of Abbreviations

BM	Business Model
BMES	Business Model Ecosystem
BMI	Business Model Innovation
BMIL	Business Model Innovation Leadership
BMIM	Business Model Management
BMrN	Business Model relation Node
BMrNP	Business Model relation node port
BOP	Bottom of the Pyramid
C	Competence
CBM	Closed Business Model
CTIF	Center for Tele for Instruction
CU	Customer and User
ERP	Economic Resource Planning System
ICI	International Center for Innovation
ICT	Information and Communication Technology
VC	Value Chain Function
WIB	Women in Business
ETSI	European Telecommunications Standards Institute
IC	Interlectual Capital
IEEE	Institute of Electrical and Electronics Engineers
MBIT	Multi Business Model Innovation and Technology
MIT	Massachusetts Institute of Technology
N	Network
NBBM	Network Based Business Model
OBM	Open Business Model
R	Relation
VP	Value Proposition
VF	Value Formula

PART 1

The Multi Business Model Approach

1

MBIT Research and Research Group: A History

Peter Lindgren

1.1 Introduction to the Background and History of MBIT

As I have mentioned elsewhere (Lindgren 2017), the research work of my PhD actually laid the ground for my interest in business models and later multi business model innovation. My son Thorbjørn and others have often asked me:

What are you working on and what is actually your job?

To explain this I thought it could be valuable to tell you briefly the story of the Multi Business Model Innovation and Technology (MBIT) research group and how I became interested in multi business model innovation.

This will also – I hope – explain why I am researching MBIT and why every morning I wake up with a smile on my face, eager to begin my work on discovering new dimensions of the manifold ecosystems of business models.

I have already covered my research before 2005 in the book *Network Based High Speed Product Development* (Lindgren 2017), so let's begin in the year 2005 when I began, together with a group of researchers, to investigate the "DNA of the business model".

The Newgibm (New Global ICT-based Business Models) project and its corresponding book (Lindgren 2011) were the very first outputs of this work. Projects like Global E-commerce and Global Innovation followed and gave more confidence and motivation to study Multi Business Model Innovation as a way to resist the increasing amount of product innovation carried out under high speed and understand why neither product innovation nor service innovation were enough for businesses to survive in the coming years.

At this time Zara Inditex had already grown very large and Ryanair was slowly taking over more and more of the low-cost carriage business model ecosystem (BMES). Large and established businesses (SAS, Lufthansa, KLM, Air France) were slowly feeling "the breath" of these new types of businesses

Figure 1.1 First things first – understand the components of the business model. Adapted from Yariv Taran's PhD (Taran 2011).

and their business models – founded and built on the Multi Business Model Approach.

However we did not as a research group quite understand how they (Ryanair, Zara Inditex and a little later Facebook, LinkedIn and Twitter) operated their businesses. Maybe, like a "humble bee", they did not even knew why they were flying themselves – but they did. We were trying to touch the "elephant" from different angels – but we had still no idea how the BM really look like and how we could define it (Figure 1.1).

However, what we could see was that the accepted and generally agreed upon best practice innovation and business innovation tools, frameworks, strategies and theories seemed not to work. This was a major trigger for us as researchers to find out how these businesses were thinking and doing business.

In August 2006 we applied to establish the International Center for Innovation (ICI) at the Center for Industrial Production at Aalborg University. The application was made as a network-based business model innovation project firstly by a core research group of six members meeting for three days in a November storm at a small fish restaurant called Niels Juel at Cold Hawaii (Cold Hawaii 2016) in Klitmøller in the Danish National Park in Thy (Pictures 1.1 and 1.2). The place itself is a case study of Multi Business Model Innovation and technology in action, and well worth a visit.

Later on, when gathering about 40 motivated network partners at the Cold Hawaii residence (Cold Hawaii 2016), the final project strategy for

Pictures 1.1 and 1.2 Niels Juel Restaurant in Klitmøller at the Danish National Park in Thy where the ideation and conceptualization of ICI took place.

Pictures 1.3 and 1.4 The strategy formation process for ICI.

ICI was produced under the process supervision of Kaj Voetmann, who also contributed later to the formation of the ICI LAB (Pictures 1.3 and 1.4).

Again the project meetings were held at Niels Juel Restaurant, hosted by its owner, Jesper Nielsen, who understood immediately what we were trying to achieve and was very happy to host us at this rural area by the North Sea.

The final core values of ICI, together with the cultures of ICI's core vision and the final application, were discussed, elaborated and finalized. The business model innovation strategy behind establishing ICI was decided upon and the operation plan was written. We could now "act and do" and make ICI happen.

After a long evaluation and acceptance phase from the EU, the Ministry of Economics and the Growth Forum for the Northern Region, in November 2007 we finally gained acceptance to fund ICI with 5.1 million euros. It enabled us to establish an applied research centre focusing on business model innovation in networks with a total budget of about 11 million euros heavily supported by industry – especially entrepreneurs and SMEs.

Picture 1.5 International Center for Innovation located at Aalborg University, Fibigerstraede 16, DK-9220, Aalborg, Denmark.

In November 2007 at the opening of the ICI (Picture 1.5) we invited some of the most well-known academics in business modelling at the time to give opening speeches on their work with business models and business model innovation (Pictures 1.6, 1.7 and 1.8).

Speeches were given by business theorist, author and consultant Alexander Osterwalder of Lausanne, Switzerland, Professor Henry Chesbrough of De Haas University, Berkeley, CA, USA and Professor Christopher L. Tucci of the College of Management of Technology, Entrepreneurship and Innovation, Lausanne, Switzerland.

In 2008 The ICI opened its first International Innovation Hub at the Innovation Center Denmark's department at Silicon Valley, CA, USA (Innovation Center Denmark, n.d.) as part of one of the seven work packages that we had promised to deliver to our funders.

The opening for this centre was held at Stanford University Faculty Club (Pictures 1.9 to 1.18), gathering figures in Danish and American industry, organizations and academia as well as entrepreneurs, to reveal the core value of ICI – research-based business model innovation – in an interdisciplinary collaboration between stakeholders from all kind of competence fields.

Later on, ICI connected with the Innovation Center Denmark's departments in Shanghai and Munich (Innovation Center Denmark, n.d.). In October 2010 a midterm evaluation report was made by COWI consulting group evaluating the first three years of ICI (COWI 2010). The evaluation was very

Pictures 1.6, 1.7 and 1.8 Business model guest speakers at the ICI opening.

positive and ICI was allowed to finish the last 2.5 years of its total funded time period. ICI finished its funded operation in spring 2013.

ICI formed the basis and platform for a young research team (Picture 1.19) focusing on business model innovation research from an interdisciplinary perspective.

Different business model (BM) typologies and frameworks, and approaches to innovating business models were studied with academic partners worldwide and businesses who laid their projects on BM and Business Model Innovation (BMI) open to study from different academic angles and viewpoints. Several researchers valued and contributed to the knowledge of BMs and BMI from this work.

Associated Professor Yariv Taran did valuable research work on verifying and finding the fundamental "building blocks" of the BM – which we later called **business model dimensions**. This work was published in his PhD Dissertation "Rethinking it All: Overcoming Obstacles to Business Model Innovation" (Taran 2011) and later on in the paper "A business model

Pictures 1.9, 1.10 and 1.11 Participants listening to speakers at the ICI opening in Silicon Valley, USA.

Pictures 1.12 and 1.13 Special guest speakers Professor Larry Leifer of Stanford Mechanical Engineering, Member of Bio-X and Affiliate of Stanford Woods Institute for the Environment, CA, USA, and Professor Woody Powell, of Stanford University Center on Philanthropy and Civil Society, CA, USA speaking at the ICI opening in Silicon Valley.

Picture 1.14 Rebecca Hwang, Head of Development at YouNoodle.com, speaking at the ICI opening at Stanford University Faculty Club.

Pictures 1.15 and 1.16 Professor Jeffrey Schnapp of Humanity Lab, Stanford University, Henrik Bennetsen, Head of Research, Humanity Lab, Stanford University and Joachim Krebs, Chief Technologist at YouNoodle.com with the company's Head of Development Rebecca Hwang at the ICI opening at Silicon Valley.

innovation typology" (Taran et al. 2015). Yariv Taran also contributed to our first work on risk related to BMI and how to classify this risk. He proposed a framework covering complexity, radicality and reach, which we will comment on later in this books Part 1 and Part 2.

Ailin Mazura Abdullah contributed to the first work on **business model innovation leadership (BMIL**) (Abdullah and Lindgren 2008; Lindgren and Abdullah 2013), especially covering the difference between BMIL and business model innovation management (**BMIM**).

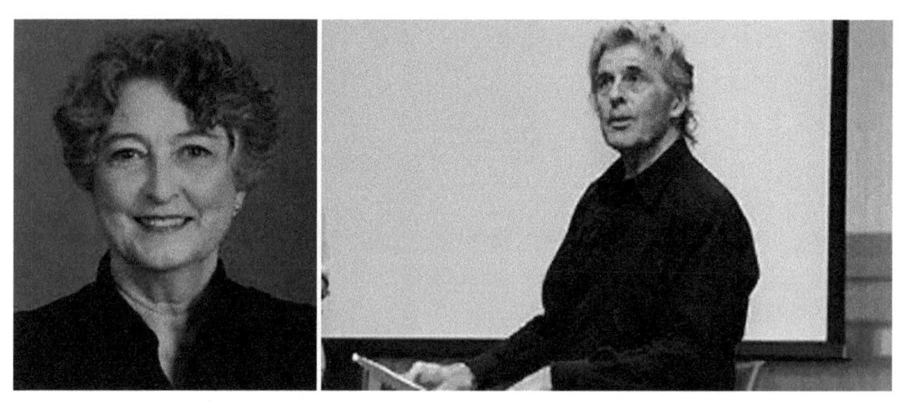

Pictures 1.17 and 1.18 Martha Russell, Executive Director of mediaX at Stanford University and Senior Research Scholar with the Human Sciences Technology Advanced Research Institute at Stanford, and Professor Keith Devlin, co-founder and Executive Director of the university's H-STAR institute, co-founder of the Stanford mediaX research network and Senior Researcher at CSLI, speakers at the ICI opening in Silicon Valley.

Picture 1.19 ICI Research Team 2008 – Professor Peter Lindgren, Associate Professor Yariv Taran, Research Fellow Ailin Mazura Abdullah, Associate Professor Rene Chester Goduscheit, Research Fellow and Consultant Jacob Høi Jørgensen, Research Fellow Subria Clemmensen, Research Fellow Kristin Falck Saghaug and Assistant Professor Jacob Ravn.

Subria Clemmensen in her very first work contributed to the concept of **Green Business Models** (Lindgren and Clemmensen 2008) and the earliest ICI research on sustainable business models.

Associate Professor René Chester Goduscheit commenced the research group's first work on **network-based business model innovation**, focusing on the basics of how businesses can lead a business model innovation project in an inter-organizational network. He contributed to the research group's earliest work on networks and how to define a network.

Jacob Høj Jørgensen contributed to our first work on **customer innovation leadership** and the difference between **customers** and **users** in business model innovation. He also contributed to seeing the value proposition of a business model (products, services and processes) from both the user's and customer's viewpoint. This work laid the first foundation for the **business model panorama view**.

Assistant Professor Jacob Ravn contributed to business models operating at the **bottom of the pyramid (BOP)**; in other words business models operating in a BMES where users or customers have no money or very small amounts of money (Ravn et al. 2009, 2010; Rayn 2012). Jacob Ravn, together with Martin Kroghstrup, also contributed to research on network-based business model innovation targeting the BOP market.

Kristin Falck Saghaug contributed to the work on **competence** – specifically the **human resource part of the BM** and **the values of the human related to BMI**. She published her PhD dissertation at the Faculty of Engineering at Aalborg University combining **theology and business model innovation** (Saghaug 2015) and made several contributions on the release of intellectual capital (IC) in BMI (Lindgren, Rasmussen and Saghaug 2013). Further, she contributed to **BMIL** related to how to **release knowledge** from SME businesses involved in BMI and to a deep research study on **BMI related to the Blue Ocean Strategy approach** (Lindgren, Saghaug and Clemmensen 2009, 2010). She also made a very important contribution on **difference related to IC and BMI** (Saghaug and Lindgren 2010).

Rasmus Joensen worked with **the establishment of the ICI** and **the ICI platform**, enabling businesses in Northern Denmark to develop and implement new global business models. His own business became one of the first partners in the project.

Rasmus Jørgensen contributed to a deeper understanding of the BMI model related to previous innovation models (Lindgren, Jørgensen et al. 2011). His work, together with his stay in Stanford University and the Innovation Center Denmark, made it much clearer to us how an innovation model for business models differs from other innovation models. His work laid the ground for proposing a sixth generation of business model innovation models (Lindgren, Jørgensen et al. 2011). Together with Kristin Margrethe and Yariv Taran we proposed and worked with new generations and ideas about BMI models. In this context Kristin Saghaug did a very valuable study on 36 women-owned businesses in Scandinavia (Norway, Sweden and Denmark) in the EU-funded project "Women in Business" (WIB 2013). This led to a new BMI process model which we will comment on in more detail later in the book.

Several businesses and not least the Region North Jutland valued this initial work on BMI. These businesses were invited in and to participate in the ICI Lab which later on laid the ground for the concept of the Multi Business Model Innovation Lab.

1.2 The Multi Business Model Innovation Lab

ICI hosted a first prototype of a **Business Model Innovation Lab**, where 11 network-based business model innovation projects were created and brought to their business model ecosystems in 2008–2013. All projects were based on minimum of five businesses from very different competence fields and included several researchers. Funding was given to the businesses to innovate their business models, but it could only be used to buy knowledge – not equipment, buildings or administrative help.

Provital, Assess2innovation, Skywatch, Space Creator, Mobile Tracking, SAFE and Cspot were some of the business models, who were established on the ICI platform. Comspace, Gabriel, COWI, Jydsk Løfteteknik, Skagen Beton, Hanstholm Havneforum, Skov A/S, Dolle A/S, Tankegangen, Skagen Foods, Acula, Aikon and many more valued the research and collaboration with ICI. All of the "new business models" were – when accepted – motivated to enter the ICI Business Model Innovation Lab as seen in a sketch model in Figure 1.2.

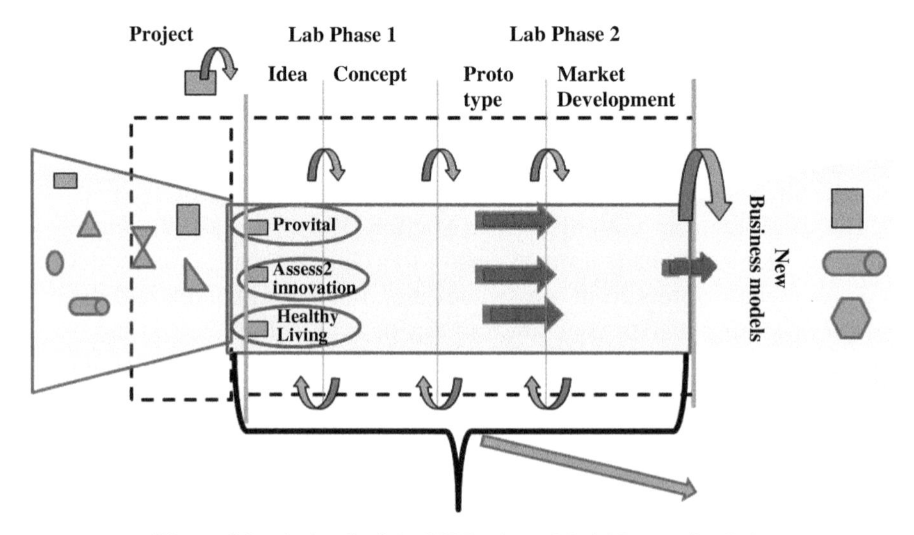

Figure 1.2 A sketch of the ICI Business Model Innovation Lab.

Pictures 1.20, 1.21 and 1.22 Professor B. J. Fogg, Stanford University, US; Margarita Quihuis – Co-Director, Peace Innovation Lab at Stanford University, Mark Nelson, Director, Peace Innovation Lab at Stanford University; Professor Morten Karnøe Søndergaard, Aalborg University.

In 2010 ICI opened a new collaboration with Stanford Peace Innovation Lab around business models and BMI on behalf of peace. The Obel Fund supported this initiative and B. J. Fogg, Mark Nelson and Margarita Quidis were central to this work, together with Professor Morten Karnøe Søndergaard, Aalborg University as seen in Pictures 1.20, 1.21 and 1.22.

This began a new focus on and contribution to our business model research in ICI, concentrating on values of business models other than money. This work, together with the Neffics (Neffics 2012) BMI case study of the HSDJ Children's Hospital in Barcelona, Spain owned by 1,200 monks, laid the ground for the change of one of the names of the seven dimensions in the Business Model Cube. The term "**value formula**" instead of previously just profit formula was chosen. It became very clear to the researchers that BMs focused and established on other values than money were existing and very highly important to understanding the game of business and multi business model innovation in the future.

At the research stay at Stanford University the interest of **persuasive technologies** (Fogg and Kaufmann 2003) and **persuasive business models** (Lindgren, Søndergaard et al. 2013) also commenced. Professor Morten Karnøe Søndergaard, Niels Einar Veirum and Katharina Wopulus contributed tremendously to the research and development in this field (Lindgren, Søndergaard et al. 2013). This research laid the vision for MBIT to be able in the future to create and study persuasive business models.

Parallel to the establishment of ICI, a new Center for Tele Infrastructure (CTIF) had been established, led by Professor Ramjee Prasad and his research team. CTIF focused on wireless and future wireless technologies, security

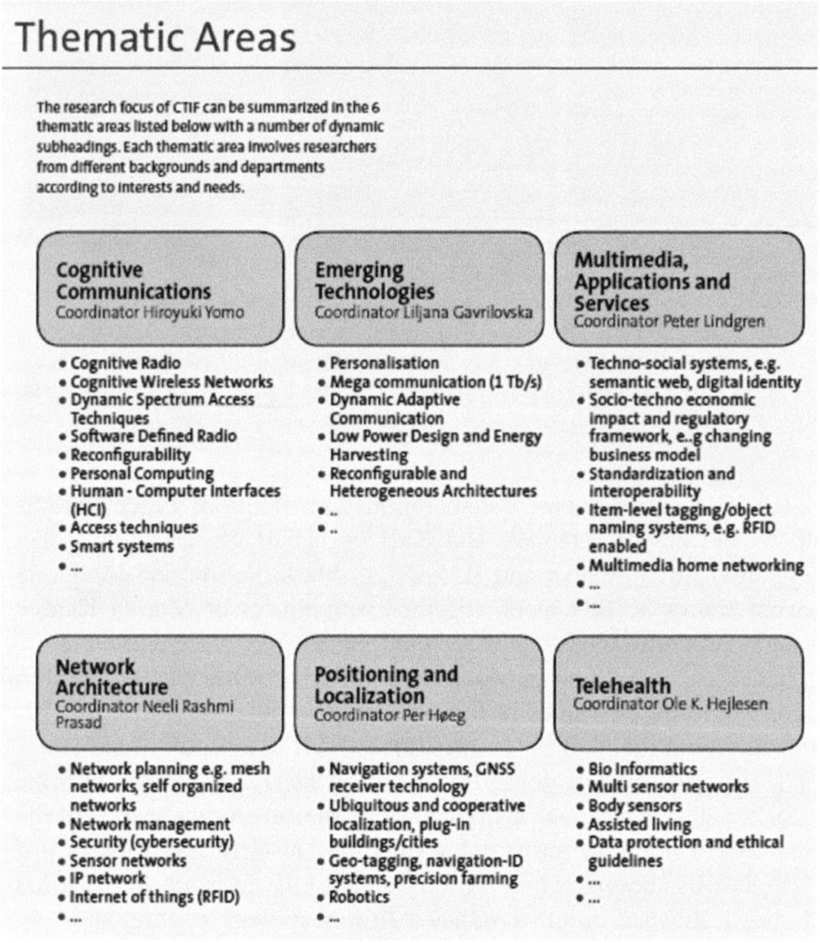

Thematic Areas

The research focus of CTIF can be summarized in the 6 thematic areas listed below with a number of dynamic subheadings. Each thematic area involves researchers from different backgrounds and departments according to interests and needs.

Cognitive Communications
Coordinator Hiroyuki Yomo

- Cognitive Radio
- Cognitive Wireless Networks
- Dynamic Spectrum Access Techniques
- Software Defined Radio
- Reconfigurability
- Personal Computing
- Human - Computer Interfaces (HCI)
- Access techniques
- Smart systems
- ...

Emerging Technologies
Coordinator Liljana Gavrilovska

- Personalisation
- Mega communication (1 Tb/s)
- Dynamic Adaptive Communication
- Low Power Design and Energy Harvesting
- Reconfigurable and Heterogeneous Architectures
- ..
- ..

Multimedia, Applications and Services
Coordinator Peter Lindgren

- Techno-social systems, e.g. semantic web, digital identity
- Socio-techno economic impact and regulatory framework, e..g changing business model
- Standardization and interoperability
- Item-level tagging/object naming systems, e.g. RFID enabled
- Multimedia home networking
- ...
- ...

Network Architecture
Coordinator Neeli Rashmi Prasad

- Network planning e.g. mesh networks, self organized networks
- Network management
- Security (cybersecurity)
- Sensor networks
- IP network
- Internet of things (RFID)
- ...

Positioning and Localization
Coordinator Per Høeg

- Navigation systems, GNSS receiver technology
- Ubiquitous and cooperative localization, plug-in buildings/cities
- Geo-tagging, navigation-ID systems, precision farming
- Robotics
- ...

Telehealth
Coordinator Ole K. Hejlesen

- Bio Informatics
- Multi sensor networks
- Body sensors
- Assisted living
- Data protection and ethical guidelines
- ...
- ...

Figure 1.3 CTIF research focus areas.

technologies and many more research fields related to these topics. Many more topics were added through the years that followed (see Figure 1.3).

We knew at that time that CTIF's research and global partners (Figure 1.4) were doing world-class research and had access to some of the best businesses in the world (Huwaii, Tata, Cisco, Nokia, Ericsson, Alcatel-Lucent, ETSI, IEEE, Princeton University, MIT and many more). It was a perfect network partner and match to ICI and to fulfill ICI's vision and strategy. CTIF had the technology and technology research. ICI was researching and developing the business model research needed and valuable for the technology.

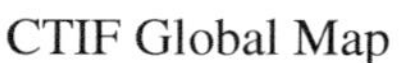

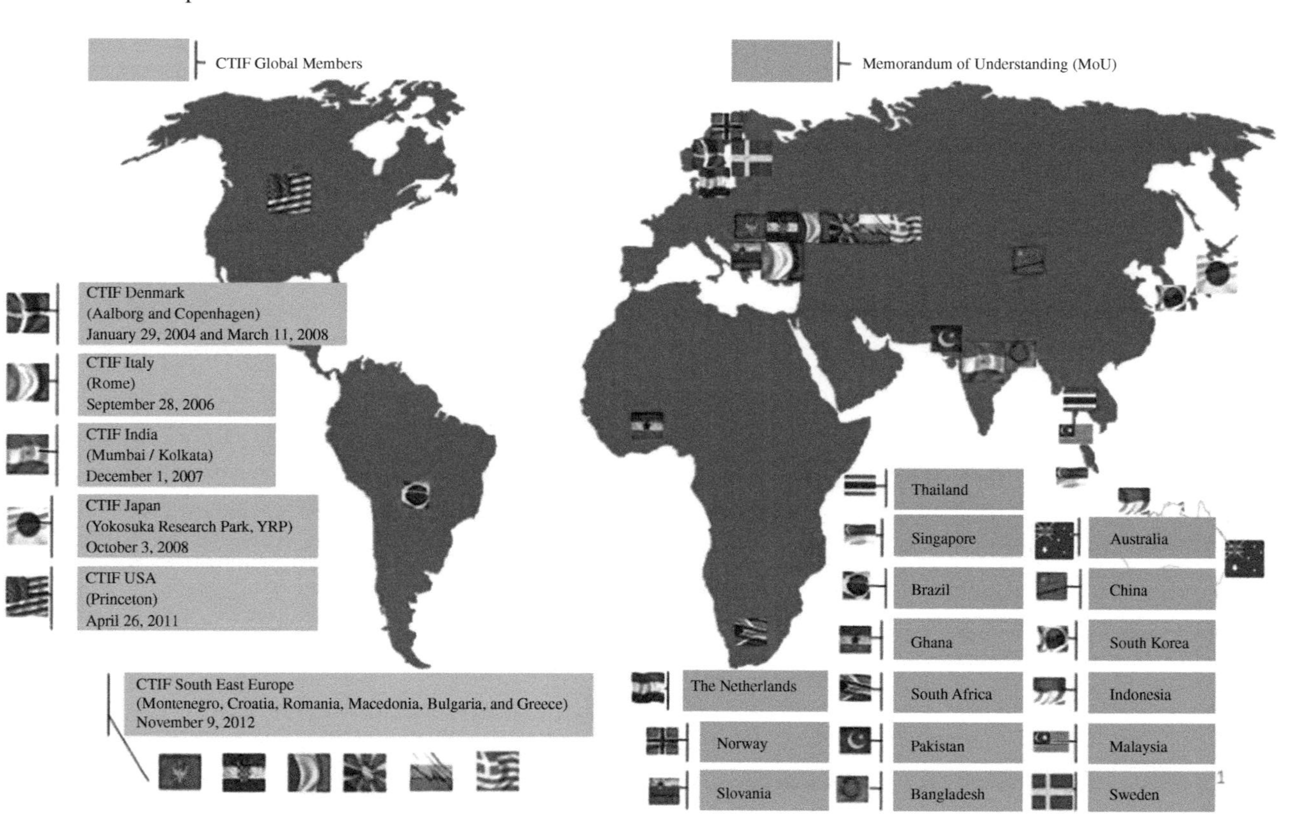

Figure 1.4 CTIF's global partners 2016.

CTIF became highly valuable to our business model research from the very first moment. In 2006, when ICI was invited to join CTIF and later made me responsible for its thematic area of Multimedia, Applications and Services we became even closer network partners.

1.3 The Establishment of the Multi Business Model Innovation and Technology (MBIT) Research Group

In 2011, when I was still working as a guest researcher at Stanford University, I took a very tough decision. I decided to leave ICI and wrote a letter to its board to announce that I would leave my post as head of the ICI research centre. I took this decision because for a long time I had felt that the rest of the management of ICI was not following the vision and strategy we originally had agreed upon, or fulfilling the promises we had given to our supporting network partners and not least our funders – EU, Ministry of Economics and Region North Jutland.

A midterm evaluation report had showed that we were on track but our new 2013–2017 strategy for ICI was not accepted in November 2010 by the board of ICI and the University of Aalborg in particular. I deeply felt that ICI was taking a direction that was not in line with its original idea, aim and application. Most important, I felt that our research on BMI was not ambitious enough to become world class and I feared that it would fall back if we could not realize our 2013–2017 proposed strategy.

So in February 2011 I left ICI and began to think of a way to realize the proposed strategy that we had had for "the ICI business". I began this journey by preparing and configuring a new research group with my old master and PhD Supervisor Associate Professor Kim Bohn, Aalborg University (Picture 1.24). Post Doc Ole Horn Rasmussen (Picture 1.23) and some master's students at the Institute of Mechanical Engineering at Aalborg University joined us very soon afterwards and in 2012 we could finally found the Multi Business Model Innovation and Technology (MBIT) research group, based on the original proposed strategy for ICI but with a different set-up.

Ole Horn Rasmussen contributed in this time period to the work on **the BM Cube** and **the relations of BMs** – specifically the concept of **the relations axiom** as a tool and framework to map the relations between BMs (Rasmussen and Lindgren 2015). Further, he contributed to **the theoretical verification of the BM Cube**, the relation axiom and the concept of the **BMES**. His work on **the reverse butterfly model** laid the theoretical ground for the connection between the BM and the BMES (Rasmussen and Lindgren 2016a).

Pictures 1.23 and 1.24 Researcher, Teacher and Consultant Ole Horn Rasmussen and Associate Professor Kim Bohn.

Pictures 1.25, 1.26 and 1.27 Head of Aarhus University B-tech Professor Michael Goodsite, Head of Department of Engineering Professor Thomas Toftegaard, Head of Aarhus University B-tech Jacob Eskildsen.

In spring 2013 Associate Professor Annabeth Aagaard and I began to work out how to realize the MBIT strategy (MBIT 2014) and proposed this to Aarhus University – Head of Aarhus University B-tech Professor Michael Goodsite (Picture 1.25) and Head of Department of Engineering Professor Thomas Toftegaard (Picture 1.26) – in summer 2013.

After some adjustments and negotiation, the new MBIT strategy 2014–2020 was accepted by Aarhus University and Associate Professor Anna Beth Aagaard and I were hired at the University of Aarhus – Department of Business Development and Technology – under our new Head of Aarhus University B-tech Jacob Eskildsen (Picture 1.27).

Two MBIT labs were initially established in Aarhus University – one in B-tech Herning (Picture 1.28), and one in the Science and Technology Navitas Building (Picture 1.29).

Picture 1.28 and 1.29 MBIT labs at Aarhus University: B-tech Herning; and Science and Technology Navitas Building.

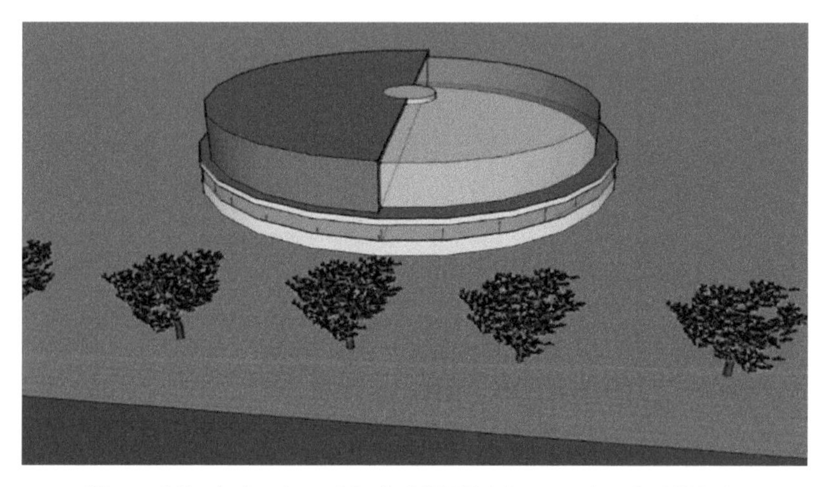

Figure 1.5 A sketch model of a MBIT lab innovated at the ICI Lab.

This was not a new idea. It had already been initiated by researchers at ICI under research assistant Gert Spender who drew up a sketch model of a MBIT lab that could be replicated to as many places as needed, both physically and digitally (see Figure 1.5).

The inspiration was very much taken from the Stanford University Clark Center (Pictures 1.30 to 1.34).

Here young students, researchers and business people meet in a building or place where they innovate future business and technology. What inspired me very much during my stay and several visits was the Clark Center's concept – built around people meeting at the restaurants and cafes at the floor level, with plenty of meeting rooms and spaces to talk and eat together, and food and drink from all over the world. The research labs are just nearby or on the next floor in a building, and here more detailed discussions and innovation

Pictures 1.30, 1.31, 1.32, 1.33 and 1.34 Stanford University Clark Center.

can take place. The Clark Center is an open centre where everybody can walk in – researchers, business people, students... It is a perfect place for BMI – and research-based BMI.

1.4 The MBIT Strategy

The MBIT strategy was built with this inspiration and as a five-year strategy – 2014–2019. From the very first moment it was an open business model strategy (MBIT 2014), inspired by Professor Henry Chesbrough's ideas and concepts, with a vision to move the MBIT Lab out in the open space – to where the BMI projects were actually happening and really taking place.

The MBIT strategy aims:

- To create a world-class interdisciplinary research centre for MBIT.
- To become an increasingly attractive research project partner for local, regional, national and international businesses, institutes and universities.
- To contribute to the engineering study programmes at B-tech (primarily at master's level, but also at bachelor's degree level) through research-based teaching and by profiling the programmes to the local business and educational communities.
- To support businesses of any kind with business model innovation and business model innovation technology, solutions and tools.
- To create a vibrant talent development environment for students at PhD level and at the final stage of the MSc in Engineering study programme.

The MBIT qualitative objectives and interests were formulated in 2013 within three areas:

1.4.1 Research

MBIT pursues and publishes research in the area of engineering and engineering management as consisting of network-based and integrated businesses and their business engineering design methodologies, which include:

- Business engineering operations frameworks
- Business technological artefacts
- Multi business model and innovation systems
- Business model information systems
- Business model information technology
- Multi business model modelling
- Entre-, intra- and interpreneurship
- Strategic multi business model innovation

- Sustainable business models
- Data-driven business models

1.4.2 Education

- MBIT contributes to the engineering study programmes at Aarhus University – B-tech, and in particular, to the research base underpinning the MSc in Engineering study programme (cand.polyt.)

1.4.3 Business

- MBIT aims to build bridges between technology and business. MBIT aims to bridges businesses of all kinds (private, public, NGO...) in means of business development, business development technologies, and entre-, intra- and interpreneurship as well as strategic business model innovation.

To pursue these themes, MBIT aims at executing research-based BMI activities such as:

- Combining business models and related big data from experiments with (applied) science to understand the processes in scientific and engineering business model challenges and problems
- Designing science research, i.e. understanding, analysing and systematizing frameworks as well as strategic approaches to determine the design and combination of business model innovation
- Mapping analytical tools to understand business models and business model innovation in businesses and in different business model ecosystems from a business innovation perspective, e.g. technological business model innovation as a prerequisite for integration and efficiency of business models, or business model competences and capabilities as enablers for different multi business model configurations
- Applying an open approach to technology, engineering and business model innovation in the context of changes and supporting diversity by combining 'as-is' and 'to-be' business models, businesses and business model ecosystems into new opportunities
- Applying business model innovation leadership and management processes with a particular focus on business model innovation leadership, strategic leadership of business model innovation and operations processes. This particularly supports complex and changing business models and business model ecosystems related to businesses and business model innovation

– Ensuring the involvement of technological and engineering aspects in concrete business model design implementation and operation of business model solutions
– Ensuring entre-, intra- and interpreneurship in business innovation
– Ensuring execution, monitoring, control, evaluation and adjustment of business model innovation

The quantitative objectives can be seen in the MBIT Strategy document (MBIT 2014).

From the very beginning, the MBIT research group has invited master's degree and PhD students to participate in MBIT research activities and projects. The model for MBIT research was sketched out from the very first draft of the strategy, as shown in Figure 1.6.

The MBIT Research Lab approach aimed at taking in different research projects within the field of MBIT's research area and attracting different competences from different faculties – art, science and technology, business, social science and health – either internally from Aarhus University or from other universities or knowledge institutions. The approach in MBIT was to work with networks of businesses – seldom one business alone – because there is and was a strong belief in MBIT Research Group that future BMs will be created in networks of different competences. This idea and concept of BMI was not new but inspired by the Newgibm (Lindgren 2011) approach that we worked with back in 2005–2007 (Figure 1.7).

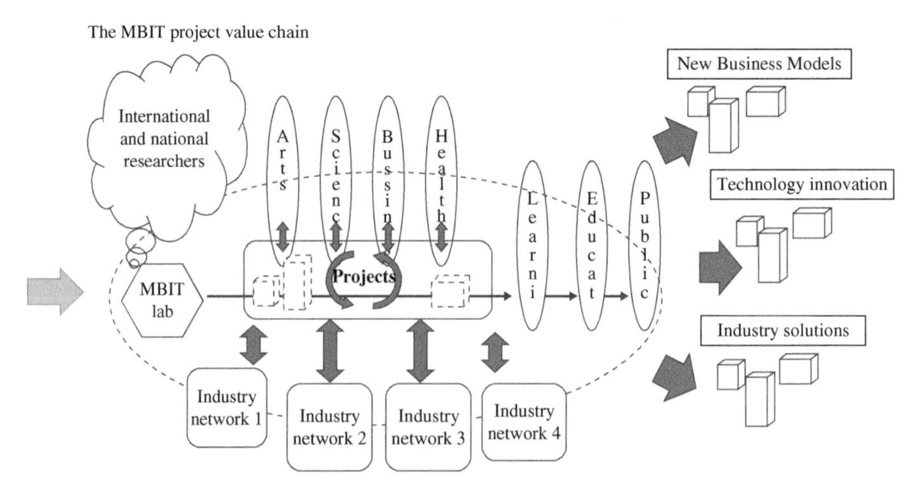

Figure 1.6 The MBIT Research Lab approach.

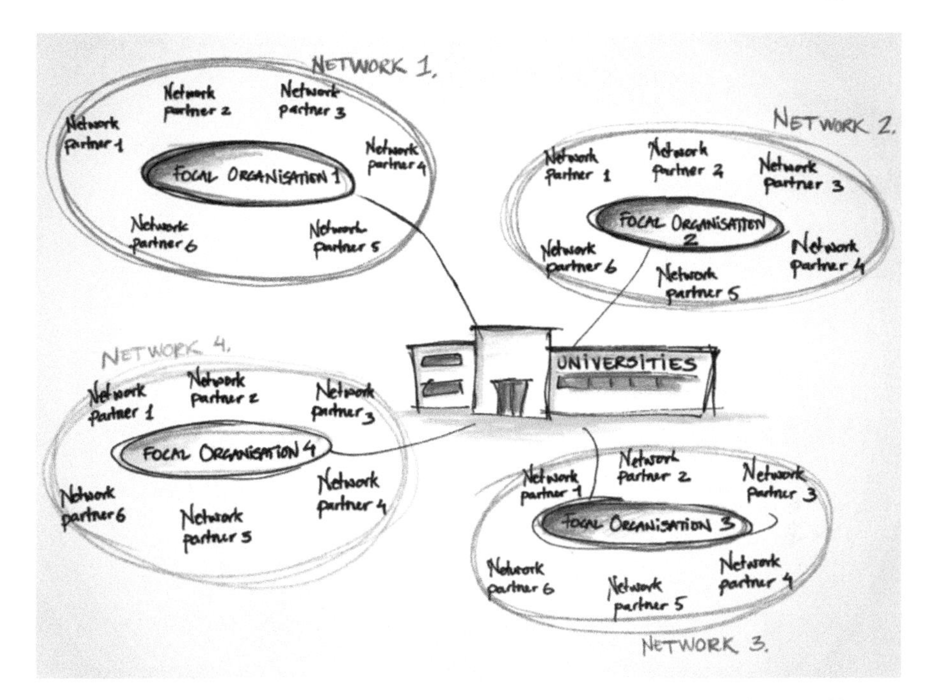

Figure 1.7 The Newgibm approach adapted from the Newgibm book (Lindgren 2011).

The very difficult task for the MBIT Lab is to make it possible for researchers from all competence and knowledge fields to study the BMI projects together and be willing to do this independent of their background and competences. However, this is one of the core competences that we want to build in the MBIT research centre and the MBIT Lab. Today – August 2017 – we have gathered a team containing a wide range of MBIT researchers (Picture 1.35).

All researchers' interest and research areas could be found at the MBIT website (MBIT 2014). The MBIT research group has visiting researchers continuously at the lab and about 10–12 master's students working with the group.

The next difficult part was to make MBIT Lab independent of place. Our vision was to have MBIT Lab "move" to the place where the BMI project was taking place – either this was a physical, a digital or a virtual place. We have come some way down this road, but have some way to go. This is our ambition to achieve in prototype in 2018.

Jesper Bandsholm, one of our first master's students, participated in the MBIT Lab and in this context contributed to the study and test of the **Bee Board** and **BMES** in several empirical cases: amongst others The Green Lab

Picture 1.35 MBIT research team, August 2017.

Skive case (Lindgren and Bandsholm 2016). This was done in a physical set-up inside and outside the MBIT Lab in Herning.

Signe Stagstrup Jensen, together with Associate Professor Jane Flarup, contributed to research on **competence related to BMI** (Flarup et al. 2016) in our search for "**The DNA of BMI**". Signe and Jane studied the competence profiles of more than 400 master's engineering students, trying to identify the DNA of BMI.

Morten Laulund and Mads Buur Sandfelt contributed to the **community-centric business model ecosystem**, especially digital BMES communities and their BMs.

Malene Rønnow contributed to the first experiments in the MBIT Lab on what was called "**the B-lab**" as a working title, **a simulated BMI environment** (Figures 1.8 to 1.10, Pictures 1.36 to 1.39). The aim of the research was to find the most optimal **BMI environment** and later optimize the BMI environment in the favour of **optimizing the "production of BMs"**.

Cosmina Radu, with Ambuj Kumar and Per Valter, continued this research with experiments rolling out and testing the B-lab from 2015 to 2017. The aim

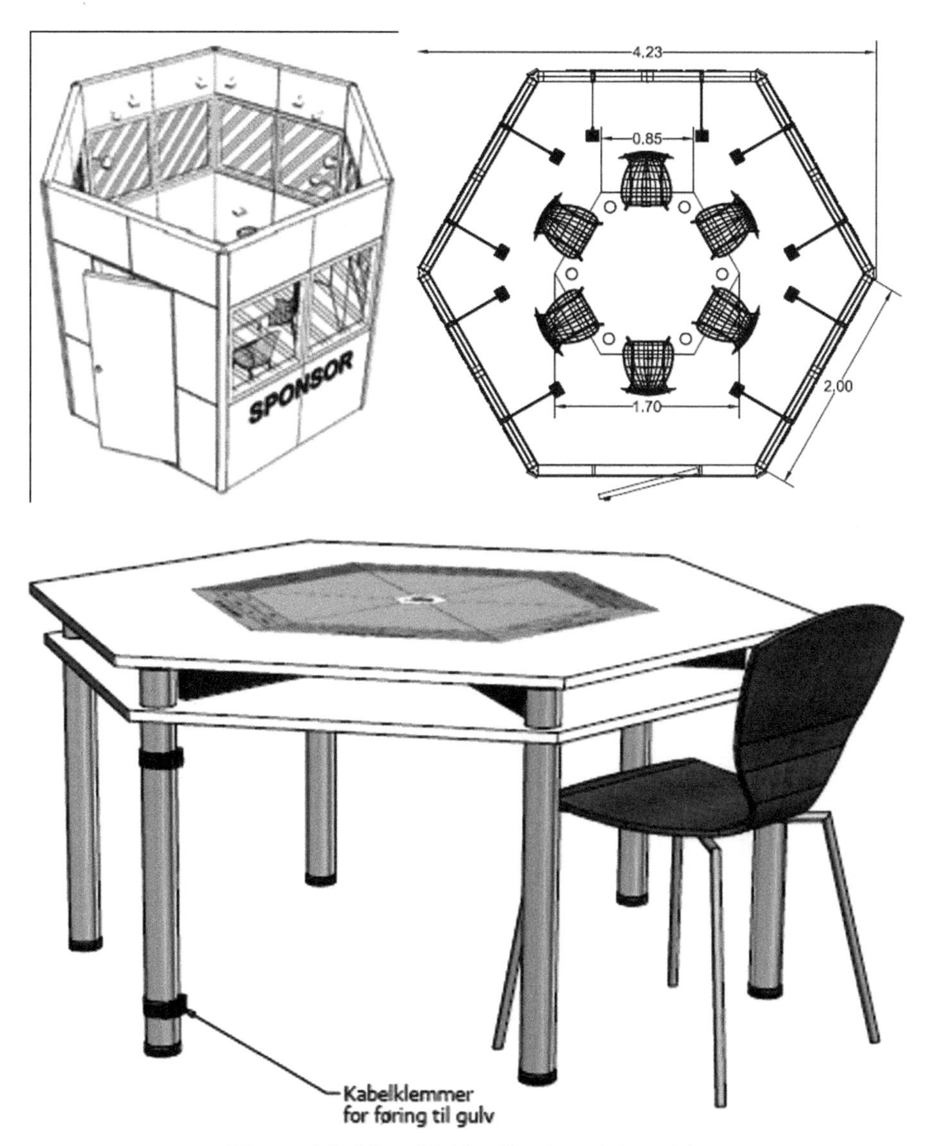

Figures 1.8, 1.9 and 1.10 Sketches of the B-lab.

was to have more than 25 B-labs in Denmark, Sweden, Norway, Bulgaria and Romania up and running by January 2018. Figure 1.11 shows a sketch of the roll out of B-labs by summer 2017.

PhD Fellow Troels Andersen currently researches and contributes to MBIT's research on strategic multi business model innovation and PhD

Pictures 1.36, 1.37, 1.38 and 1.39 Malene Rønnow together with test equipment, testing and monitoring situations in the BMI environment.

Fellow Torben Cæsar Bjerrum researches and contributes on data-based business models for the MBIT research group. PhD Fellow Kristian Løbner researches and contributes on business model innovation leadership in incumbent project-based organizations for the MBIT research group. They are all expected to finish their PhDs before 2020.

1.5 Future of MBIT Research and Research Group

The MBIT research group was in August 2017 standing close to "the start of 2018". We now had about 25 researchers in our team and had integrated Stanford Peace Innovation Lab and the CTIF Global Capsule into the lab and research group. 17 B-labs/cubes (Figure 1.11) were installed in Sweden and Denmark.

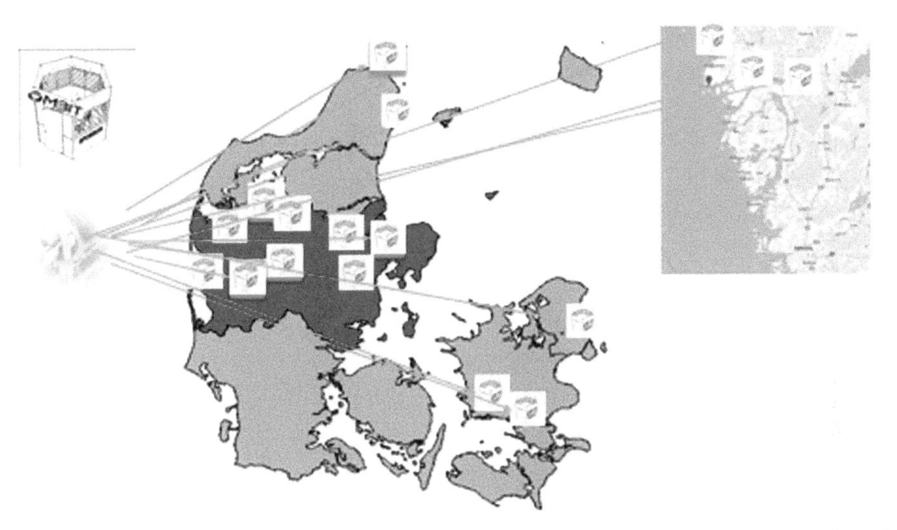

Figure 1.11 Map of the B-labs and MBIT test environment spread out in Denmark and Sweden in summer 2017.

A new strategy for MBIT 2020–2025 was under construction already started to be formulated in 24th May 2017. A first draft of the new MBIT strategy was suposed to be presented at a two-day strategy workshop hosted by one of our partners in autumn 2017. Then Head of Department Anders Frederiksen called me the 18th of August 2017 to discuss a proposed change in the setup and strategy of MBIT. This I will comment on in the Part 2 of the book.

With this introduction, which answered some basic questions on why I am working with business models and the Multi Business Model Approach, I would like to thank all members of my research team, all partners and funders that have contributed and supported ICI and MBIT in making our strategies happen. It has not been an easy journey – or an easy business model(s) – to implement the BMES. Every day we had to work hard in the field between business, technology and research. Some would say it is "blood, sweat and tears" – but it was worth it. In fact, when I turn up in the new MBIT Lab I see the same smiles and enjoyment in my research team as I saw right from the beginning of ICI. Multi business model innovation is a mindset and a language – once you have touched and learned it you will never forget.

It's like "being your business."

That is what I discovered in my many travels to businesses around the world to research and gather cases. Those business and their employees that try hard every day to "be their business" are most often those who achieve the highest performance – either measured in money or other values, or both. Often it is the other values that make them wake up early in the morning and start working.

Chapter 1 ends with giving a big thanks to Junko Nakajima and her team at River Publishers for working with me on this book. Thanks to River Publishers – and Rajeev Prasad for letting me publish this work through your business.

2

The History of the Business Model

Peter Lindgren

Abstract

The chapter gives a historical review of the business model literature. The purpose of this chapter is to describe how different academics have thought about the business model (BM) through history. A very detailed story of the business model literature can be found in many of the sources refered to in the chapter.

Previous BM concepts and related academic ideas on what a BM looks like are compared.

2.1 Introduction to the Business Model Approach

The first discussion on BMs can be traced back to an academic article in 1957. However, the concept did not gain acceptance until the mid-1990s (Fielt 2011). In Figure 2.1 an overview is given of some of the important contributions and developments in BM literature since the mid-1990s.

The question "What is a BM?" has been raised, discussed and answered by many researchers in the last decade (Fielt 2011). Porter argued that a "definition of a BM is murky at best. Most often, it seems to refer to a loose conception of how a business does business and generates revenue" (2001, p. 73). Morris et al. (2003), after reviewing existing theory on business models during the late 1990s to 2003, concluded that a business's potential creation of value cannot be explained from the BM model theory, and that "a general accepted definition has not yet emerged" (p. 8; see also Fielt 2011). However, Osterwalder et al. (2005) summed up academic work on BMs from the past 20 years, and stated that a definition of a BM broadly related to a blueprint of how an organization should conduct its business (Osterwalder et al. 2005). They further argue that a BM is a set of elements which can be referred to as building blocks that, by their interrelation, express the logic of how a business earns money (Osterwalder et al. 2005).

Many academics have, in the past, been widely recognized for their approach to the BM concept (Fielt 2011). Important to note is the distinction

The Evolution of Business Model literature

Activity	Define and classify Business Models	List Business Model components	Describe Business Model components	Model Business Model components	Apply business model concept	Define and classify multi business model concept	Define and classify Business model ecosystem concept
Outcome	Definitions and taxonomies	"shopping list" of components	Components as building blocks	Reference model and ontologies	Applications and conceptual tools	Definition and taxonomies	Definition and taxonomies
Authors	Rappa (2001), Timmers (1998)	(Linder & Cantrell 2001), (Magretta, 2002) (Amit & Zott 2001)	(Afuah & Tucci 2001), (Afuah & Tucci 2003) (Hammel 2000) (Weill & Vitale 2001)	(Gordijin 2002), (Osterwalder & Pigneour 2004) (Lindgren, et al., 2011) (Johnson, 2010)	(Osterwalder, et al., 2010) (Lindgren, 2011)	(Casadesus-Masanell R.,Ricar J.E. 2010, Lindgren 2012)	(Lindgren 2012)

Inspired and developed on behalf of Osterwalder, Pigneur and Tucci 2004)

- **'Taxonomies'** – which business models resemble each other?
- **'Reference Models'** – an abstract template for the development of more specific models in a given domain, and allows for comparison between complying models
- Organized according to some underlying common building blocks.
- **'Ontology's'** – seeks to describe or hypothesize basic categories relationships within the BM framework.
- **'Applications'** – BM's are *stories that* describing *how enterprises work.*
- **'Conceptual tools'** – BM building blocks are both physical and conceptual elements ('Meta-Model').

Figure 2.1 The evolution of business model literature up to 2012.

between business (Abell 1980) and BMs, as a business is considered in our framework to have one or more BMs, i.e. the multi business model approach (Lindgren 2012). Furthermore, all BMs can be referred to either as "as-is" BM – already operating in the BMES – or "to-be" BM – being innovated or preparing to be introduced into the BMES (Lindgren 2012).

From its infancy until today, it can be documented that the BM concept has naturally evolved and changed in relation to the BM context (Zott et al. 2010). Globalization and the internet have increased businesses' interdependency and today businesses are connected in physical, digital and virtual networks (Choi published in Turban 2003; Daft 2010; Peng 2010). Thereby, it is possible to utilize competences across businesses' BM and BM boundaries in order to strengthen the BMI (Daft 2010; Lindgren 2012) of businesses. This tendency can be argued to have influenced the BM literature. For example Chesbrough (2007) suggests that BMs should be open (Open Business Model (OBM)), so that businesses can utilize the dimensions and components of BMs of other businesses within their own BMs.

It has been argued that until 2007 the BM literature primarily concerned closed BMs (CBMs), where BMs were bound to the focal business and thereby not open to other businesses (Lindgren 2011). The CBM argued by Chesbrough (2007) was not deemed to fit in the global business model ecosystem (BMES) (Lindgren 2016b), which requires openness and interfaces being able to comprehend interfacing with other businesses' BMs. Chesbrough (2007) further claims that CBMs delimit the potential value and effective use of BMI. BMI refers to the reinvention of current BMs' dimensions or creation of new dimensions in order to create advantages to the business. Thus, Chesbrough's (2007) way of thinking of BMIs, as being open, has become the foundation of the development of a new and open network-based BM innovation concept (see also Daft 2010; Lindgren 2011). BMs are becoming more dynamic in their construction and today's BMs may easily be outdated tomorrow. Lindgren (2011) suggests that new BMs should serve as platforms for continuous BMI – and development of other BMs. Any business model is proposed as a platform for other BMs and BMI – and thereby the development of a multitude of BMs.

2.2 The Background of the Business Model Approach

Today, the term "business model" is used every day by those in business and by business model academics. Even national governments (including the US government) and the European Commission use the term "business model". The increased awareness of BMs (Casadesus-Masanell and Ricart 2010;

Teece 2010; Zott et al. 2010; Kremar 2011) have intensified the search for a generic business model language. However, with increased use of and research in BMs the fuzziness on how the BM really is constructed has increased even more.

The focus on being first with a generic and commonly accepted BM language has increased drastically in recent years (Taran 2009; Zott et al. 2010; Fielt 2011). The emphasis on the BM's dimensions has been the topic of much academic work (Magretta 2002; Osterwalder and Pigneur 2002; Johnson et al. 2008; Chesbrough 2010; Kremar 2011; Osterwalder 2011). Many theorists have focused on the question of how many dimensions the BM really consists of. Some propose four, while others propose six, nine and twelve dimensions. This raises the question of how a business model is really constructed and whether we will ever be able to find the generic dimensions and construction of the BM. Further, can we distinguish one BM's construction from another or are they really built around the same generic dimensions?

These questions imply the increasing importance of thoroughly knowing and finding the dimensions of the BM. They are also related to the question of when can we talk about a new BM and its incremental and/or radical changes (Peng 2010; Lindgren 2011), and whether that influences the generic construction of the BM.

The focus is therefore primarily on the dimensions and construction of any BM, although this is no longer deemed sufficient to cover the whole BM theory framework as it is just one focus of many – a fragmented part of the whole business model environment, research and discussion. Today, the focus of the BM seems to be changing towards a more holistic discussion, taking in the BM's relations to other BMs and the BM's Ecosystem – leaving the basic BM dimensions and constructions behind. The focus of the OBM (Chesbrough 2007; Daft 2010) and the innovation of BMs (Osterwalder 2011) seems to have taken nearly all research attention.

In an ever-changing and increasingly global competition, which according to Friedman (2007) is a result of the ongoing process of globalization and business model change, Chesbrough (2007) emphasizes the need for even more BMIs, including developing open and different businesses models. But how can a business follow this advice without knowing the basic construction of the BM? As the basis of any BM discussion we must begin by understanding, defining and testing the generic construction of the BM – what we could call the generic dimensions and questions of a BM, as shown in Table 2.1 as our proposal to these.

In Part 2 of the book we will take up some of the new evolutions and trends in academic business model and business model innovation literature.

Table 2.1 The generic dimensions and questions of a BM

Dimensions in the BM (physical, digital and virtual)	Core questions related to BM dimensions
Value proposition/s (products, services and processes) that the business offers	What are our value propositions?
Customer/s and users (users, customers, target users and customers, market segments that the business serves – geographies as well as physical, digital, virtual)	Who do we serve?
Value chain (internal) configuration	What value chain functions do we provide?
Competences (assets, processes and activities, e.g. technologies, human resources, systems and culture) that translate business's inputs into value for customers, users and networks (outputs)	What are our competences?
Network: network and network partners (strategic partners, suppliers and others)	What are our networks?
Relation/s (e.g. physical, digital and virtual relations, personal, tangible and intangible)	What are our relations?
Value formula (profit formulae and other value formulae)	What are our value formulae?

3

Comparing Business Model Frameworks

Peter Lindgren

Abstract

The focus in this chapter is primarily on the dimensions and construction of the proposed frameworks of business models (BMs). BM frameworks have been a central part of the business model community's research and discussion for many years. BM frameworks have been paid a great deal of attention in the academic business model community – however, nobody has found the generic BM framework or empirically proved one.

The BM Canvas by Alexander Osterwalder (Osterwalder 2011) is still the most well-known BM framework worldwide. But other frameworks have been proposed and new ones are emerging. In this chapter, I try to discuss some of the most well-known BM frameworks and bridge them to each other. The aim is to find BM constructions and dimensions that everybody seems to acknowledge. I also try to discuss and find those BM dimensions that distinguish BM frameworks from each other, overlap and point to dimensions that seems to be lacking.

3.1 Introduction

As the amount of literature concerning business models (BMs) has increased in recent years (Teece 2010; Zott et al. 2010; Kremar 2011) a definition and a generic framework – and some say a language – of the BM have been much needed. Nobody can explain why it is so difficult to find the generic framework – and why this search has been so long underway.

However, many can understand why academia in the business model community cannot agree. Of course it would be tremendously prestigious to be the father or mother of the BM language or framework.

However, contrary to how research in the healthcare and technology fields of science are carried out, many of the existing proposed BM frameworks and languages are not empirically tested. They are just BM framework concepts and languages that would doubly function if they ever were "implanted" to "the patient" – the business.

All this conceptualization has led to a large variety of definitions in scholarly and practical literature (Magretta 2002; Chesbrough 2007; Johnson et al. 2008; Osterwalder 2011; Gassmann et al. 2012). However, none empirically proves their own framework.

A commonly accepted generic language and framework of the BM, therefore, has been and is much needed. For many years it has been needed to embrace the opportunities but also the challenges of business models and business model innovation (BMI). A commonly accepted BM language would enable BM research to take one step towards becoming an accepted academic theory. In Table 3.1 we point to some of the advantages in having a commonly agreed upon BM language and framework.

3.2 Comparing Different BM Frameworks and Languages

In our study that began in 2006 in the ICI research group and continued later in the MBIT group, we began carefully "bridging" BM frameworks from different business model frameworks to each other as can be seen in Table 3.2.

In 2011 ICI had tried to "bridge" some of the most developed and acknowledged BM frameworks (Osterwalder's Business Model Canvas (2011); Johnson, Christensen and Kagermann's BM framework (2008); Chesbrough's BM (2007); and many more models and frameworks) to the BM Cube concept.

This research work was carried out within the EU Horizon 2020 project – Neffics (Neffics 2012) and a part of the result is shown in more detail in Figure 3.1, for example the BM canvas framework model (Osterwalder 2011) and the Johnson et al. framework model (Johnson et al. 2008).

As a result of this work we found generic BM dimensions that most theorists seemed to acknowledge – in particular, that a BM has value propositions or value offerings, that a BM has customers, that a BM has key functions, processes or activities that it carries out and that it uses key resources or competences. In Table 3.3 we map those dimensions that we found were most agreed upon, were missing and that there was some confusion around.

In Chapter 4 we discuss why we added some dimensions to the BM that our research found were missing, especially **users, relations, value chain functions (secondary functions), competences (technology, organizational systems, culture) and value formula (other values)**.

Based on our research we also discuss why we believe that some BM frameworks are too complicated, have overlaping dimensions and therefore

Table 3.1 Over all benefit categories of a common accepted and agreed upon BM language

Overall benefit categories	Benefits in detail
Interoperability in BMI	- Ability of devices and BMs to work and innovate together relied on BMs complying with standard language of BM
Support of government policies and legislation in BMI	- Standards, IPR and Patents of BMs could play e.g. a central role in the global and regional BMES policy. Standards, IPR and Patents are frequently referenced by regulators and legislators for protecting user and business interests, and to support government policies
Increase in interdisciplinary BMling across vertical and horizontal BMES	- Increase in interdisciplinary BMing across vertical and horizontal BMES due to possibility to "talk" together across BMES, Businesses, BM and thereby competences and background
Increase in BMI Technology development	- Would provide a solid foundation upon which to innovate new BMI technologies, new learning and new knowledge on BM and BMI to enhance and advance existing BMI practices
Provide economies of scale in BMI	- Would provide business to being able to "produce" and "innovate" "large bats" and invest in "mass production" of BM's
Encourage BMI and more BMI	- Standards provides business with developing BMI further on behalf of standards
Increase awareness of technical developments and initiatives within BMI and BMI technologies	- Provides platform for increasing awareness - Would provide a greater variety of accessible BMs to consumers
User, Consumer, network and "things" choice of BM and BMI would be easier to adapt	- Provide the foundation for new features and options, thus contributing to the enhancement of daily BMI – user-driven BMI, interdisciplinary BMI
Safety and reliability in BMI	- Would help ensure safety, reliability and business care. As a result, users, customers, network, competences and businesses in general would perceive standardized BM language as more dependable – this in turn would raise these stakeholders confidence, increasing sales and the take-up of new technologies and business models for BMI
Advance BMI	- Would provide a solid foundation upon for research, learning and new knowledge on BM and BMI to enhance and advance existing BMI practices

In essence, if a common or standard BM language – or a standard BM language – was accepted, present and agreed upon it would amongst others be possible to gain many benefits of BM and BMI.

Table 3.2 Business model dimensions and components

Source	Specific dimensions and components	Number of BM dimensions	Empirical support Y/N	E-commerce (E)/ general (G)/other (O)
Abell (1980)	Customer function, customer group, customer technology	3	Y	G
Porter (1985)	Suppliers, buyers, competitors, new entrance, substitutes	5	Y	G
Porter (1985)	Value chain activities – primary and support activities		Y	O
Sanchez (1996, 2000, 2001)	Product, process, technology, market, organizations, knowledge architecture	6	Y	G
Morris et al. (2003)	Value offering factors, market/customers factors, internal capabilities factors, competitive factors, economic factors, personal/investor factors	6	Y	G
Von Hippel (2005)	Users		Y	O
Goldman, Nagel and Price (1995)	Network, competitors		Y	O
Vervest et al. (2005)	Network		Y	O
Prahalad and Hamel (2005)	Competences		Y	O
Chesbrough (2007)		6	Y	G
Johnson et al. (2008)	Value, customers	4	N	G
Casadesus-Masanell and Ricart (2009)			Y	G
Casadesus-Masanell and Ricart (2010)	Value creation, value delivery and value capturing	3	Y	G

(Continued)

Table 3.2 (Continued)

Source	Specific dimensions and components	Number of BM dimensions	Empirical support Y/N	E-commerce (E)/ general (G)/other (O)
Zott, Amit and Massa (2010)			Y	G
Fielt (2011)			Y	G
Lindgren, Jørgensen et al. (2011)	Value proposition, customers, profit formula	7	Y	G
Porter and Kramer (2011)	Values, customer, supplier		Y	G
Gassmann et al. (2012)	Value creation, value delivery and value capturing. Adapted by the elements of a business model (Casadesus-Masanell and Ricart 2010).	3	Y	E/G
Lindgren and Rasmusssen (2013)	Value proposition, user/customers, value chain functions, competences, network, value formula, relations	7	Y	G
Baden-Fuller (2015)	"Who is the customer?", "What is the value created for that customer in his or her interaction with the firm?" "How is that value to be monetized (directly or indirectly)?" (see Teece 2010; Baden-Fuller and Haefliger 2013; Baden-Fuller and Mangematin 2013).	3	Y	G

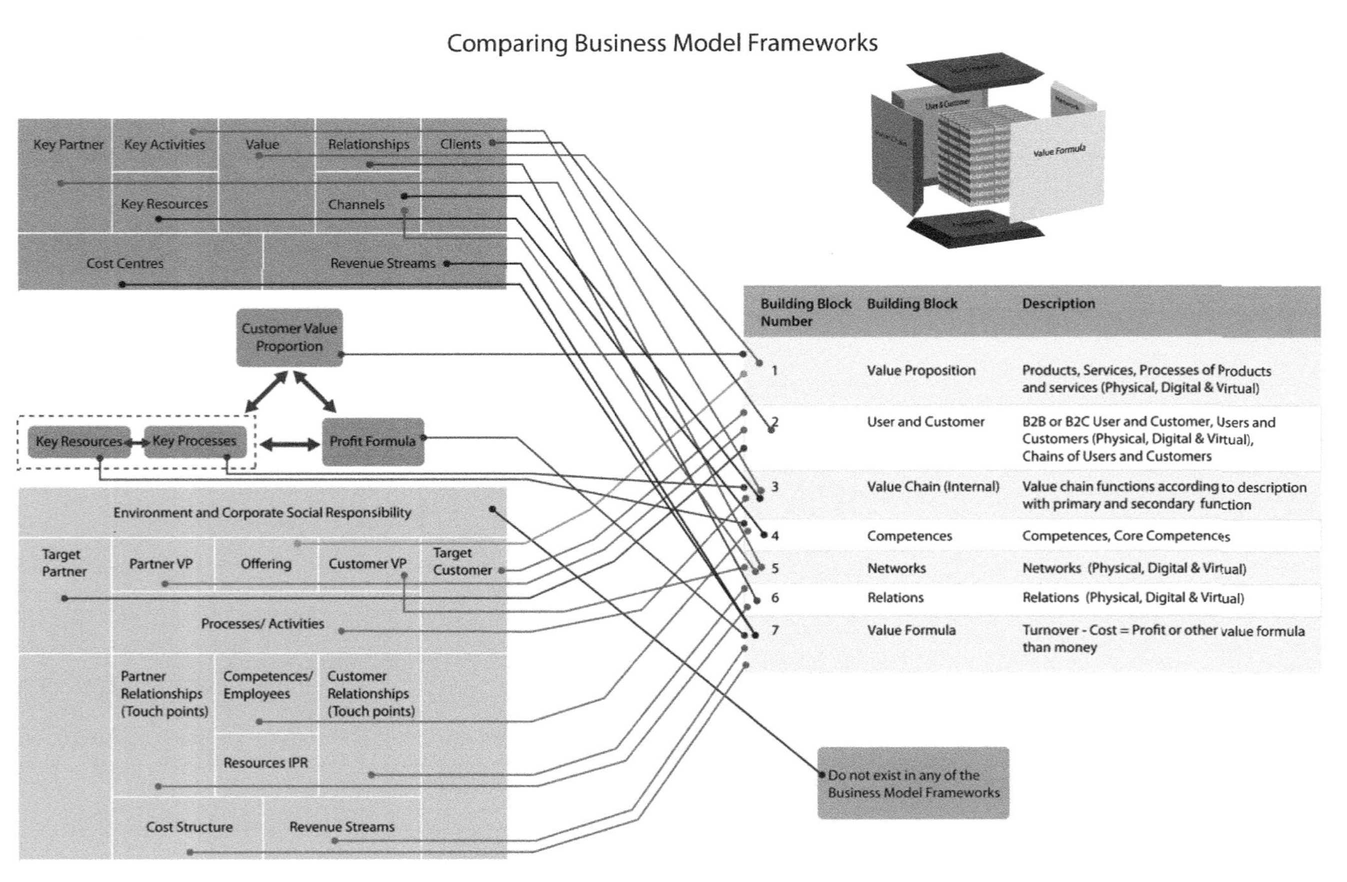

Figure 3.1 Comparing three business model frameworks to the business model cube dimensions.

Table 3.3 Business model dimensions agreed upon, missing or not defined clearly

BM dimension/s	Agreed upon	Missing	Confusion
Value proposition	Products	Several BM frameworks miss the service and process of the value proposition (process of product and services offered). Several miss digital and virtual value propositions and do not consider the integration of physical, digital and virtual value proposition	Some BM frameworks talk about values of a BM as a core value of the business but this is not equal to the value proposed to the user or customer
Users and customers	Most frameworks have customers	Hardly any frameworks consider users as a part of the BMs or relevant to the BMs	
Value chain functions	Most frameworks include activities to be carried out	Most frameworks lack the focus on primary and secondary functions or activities to be carried out	Some BM frameworks talk about activities or value chain functions as related to a value chain of suppliers, manufactures and customers
Competence and capability		Most BM frameworks do not cover the organizational systems and culture of BMs	
Network			Network partners
Relations			
Value formula	Most BM frameworks cover profit formula	Many BM frameworks do not include profit formula and most BM frameworks do not consider other values than money	Revenue stream, cost structure
"As-is" and "to-be" BMs	Most BMs' frameworks work with the operating BM	Most BM frameworks do not operate with "to-be" BMs. Most BM frameworks work with "to-be" BMs on an ideation and conceptualization basis but do not consider how to prototype and implement these	

need to merge some dimensions. We found that some BM frameworks had BM dimensions that were overlapping – e.g. in Osterwalder's framework we believe cost structure and revenue stream could with advantage be merged to one BM dimension (Taran 2011). According to other academic frameworks, a profit formula explains very well a BM's calculation method to ascertain price and costs. Further, we found that it is not the revenue stream and the cost structure that are essential for the BM to operate – it is "the calculation formula" that is vital.

In our research we also found that some dimensions proposed in different frameworks had to be taken out because they were not vital for an operating BM. We found that they were not really present and not really necessary to operate a BM or allow a BM to operate. This is probably due to the fact that our approach was very much focused on the micro dimensions and components of the BM and that we left the macro dimensions to the BMES. Therefore, we left out environmental and corporate social responsibility; further, we took out strategy, as we relate this to BMI and especially the "sensing" part of BMI.

We comment on those dimensions and terms that we found were confusing or not clearly defined. Cost structure and revenue stream are, for example, more a result of an analysis of costs and revenue but not really something that is vital for a BM to operate. A BM can have a certain cost structure but that does not mean that it will operate, or not operate. We relate this to the Ryanair BM example. Everybody knows that it is impossible cost structure-wise to fly a passenger from London to Athens for one euro. This does not mean that the BM is not operating and cannot operate. In fact it does. In other words the "BEE" or the BM "flies", but seen from a cost structure and revenue stream perspective it should not even be able to "take off".

We tried further to leave out words and classification of dimensions and components such as "partner", "target" and "key" in our framework. We found that these words are confusing and signal a strategy decision or classification that when one studies the BM carefully might not have anything to do with what and how the BM is really operating. A business might have a key partner or a target customer, but in fact the BM does not or may not even involve these in its BM operation.

We acknowledge the latest development of BM frameworks – the process BM view (Casadesus-Masanell and Ricart 2010). However, we relate the process view to the BMI process and, further, to the value proposition process that all business and business models must take into consideration. We very much agree that BMI and businesses in the future must focus more on the process view – and leave, for example, the focus of a product and service. It is the value proposition process, for instance, that is important and critical to the

Picture 3.1 A full BM value circle in a very simple market.

customer, the network and even the employee. However, we have augmented the process view (Casadesus-Masanell and Ricart 2010) with the receiving and consumption parts, so it includes and completes **the total value process for a BM**:

Create – Capture – Deliver – Receive – Consume

If a BM cannot ensure, or a business is not aware of, the entire value process for a BM then the BM will not work as intended and the value proposition will maybe never reach and be consumed by the customer. Further, the BM will not receive any value back from, for example, the customer, and the BM process will thereby not be fulfilled – which is critical for the BM and also to classify a sale as finalized (Kotler 2004). Kotler says in that case there is no market and no business. In Picture 3.1 we show how the full value circle can look in a very simple BM context. Value created, captured, delivered, received and consumed.

In this case example the buyer receives value in the form of a product and the seller receives value in form of money. We will discuss this further in the following chapters because this may be too simple a way to consider and work with BM theory.

The result of our long research work, with numerous BM cases and businesses, resulted in the proposal of a generic BM framework that we called "The Business Model Cube". We explore this in Chapter 4.

4

The Business Model Cube

Peter Lindgren and Ole Horn Rasmussen

Abstract

The Business Model Cube was developed as an output of the work mentioned in Chapters 1 to 3 and several years of BM research and empirical BM cube testing. Several researchers contributed to the formation and verification of the BM Cube and its related dimensions and components. Associate Professor Yariv Taran, in particular, contributed to the hypothetical concept of the BM with seven dimensions. Sigitas Pleikys contributed to the cube framework and digital visualization of the BM Cube (Figure 4.1).

Ole Horn Rasmussen contributed to the relations axiom framework, which will be covered in detail in Chapter 7. This chapter explains in detail the arguments of how and why the BM Cube could be a proposal for a generic BM framework and BM language. Further, it shows case examples of the use of the BM Cube in the different businesses we have studied.

4.1 Dimensions, Concepts and Language of a Business Model

The term "business" has been defined by reputed academics from several viewpoints and dimensions. Abell (1980) defined a business by just three dimensions – **customer functions (what)** (values); **customer groups (who)** (customers); and **customer technology (how)** (production technologies and process technologies) (Figure 4.2).

So, interestingly, Abell had already indicated in 1980 a cube which formed the "borders" of a business – in three dimensions, however Porter (1985) argued that a business should be defined by its **suppliers, buyers** (customers) and **value chain activities**. Hamel and Prahalad (1994) argued that a business could be defined by its **competences** and its **core competences**. Vervest et al. (2005) argued that a business could be defined by its **network** and how it organized its business together with network partners, and Johnson

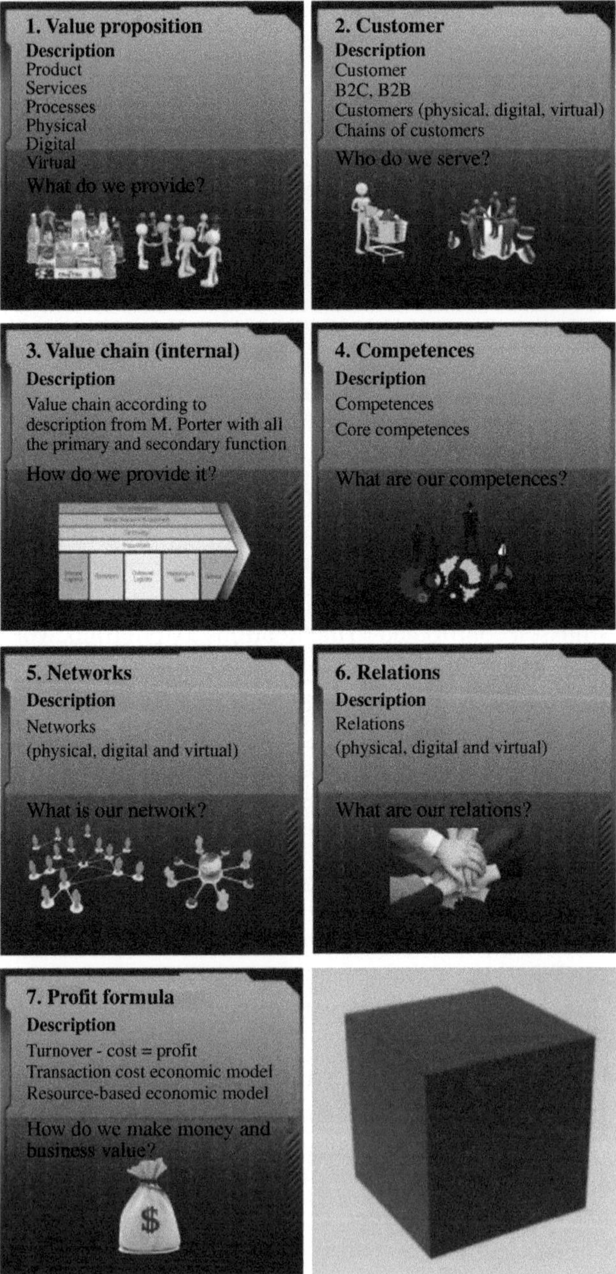

Figure 4.1 Sigitas Pleikys' first sketch of the BM Cube for our research work (Pleikys 2012).

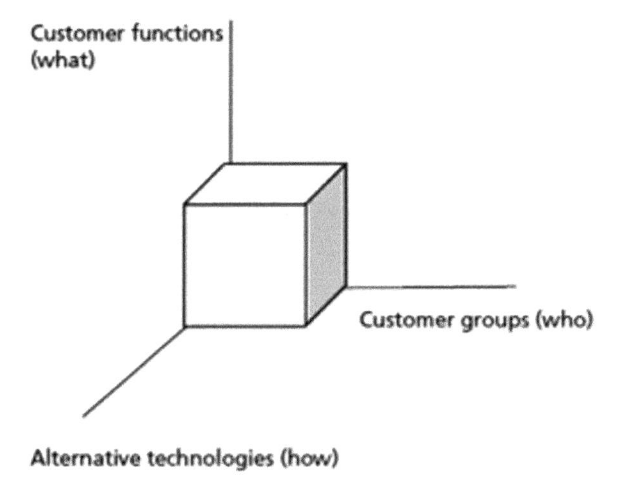

Figure 4.2 Derek F. Abell's three-dimensional business model.

et al. (2008) defined the business as how it created **value** to the **customer**. Håkansson (1980; see also Amidon 2008; Allee and Schwabe 2011; Russell 2011) defined the business by its **relations**. **Profit maximization** has been the central assumption in business and managerial economics (Henry and Haynes 1978) and the reason for the stress on profits has been that it is the one pervasive objective running through all businesses; other objectives, according to Henry and Haynes, have been more a matter of personal taste or of social conditioning and were variable from business to business, society to society, and time to time. The survival of a business has until today very much been considered as depending upon its ability to earn profits, where profits have been the business measure of its success (Henry and Haynes 1978). Another reason for emphasizing profits is their convenience of analysis and because it is easy to construct **formulae** on the assumption of profit maximization. It has been much more difficult to build models based on multiplicity of **value formulae**, especially when these formulae are concerned with non-monetary factors as "fair", the improvement of public relations and, for example, the maintenance of a customer's satisfaction. However, other value formulae than profit formulae have become very popular these days to business – even more popular than profit – especially as a reaction to, for example, the financial crisis of 2008 and global warming.

From these acknowledged academic works, we found after five years' intense research in the ICI and MBIT research groups some generic dimensions that support the idea that any business could be defined by them.

More than 12 researchers were involved in this research in the 2007–2013 time period.

From this point of entry, we then tested our BM dimensions in more than 400 different businesses to verify empirically our hypotheses of the existence of what we found were seven dimensions of any BM. This resulted in the creation and capturing of the **Business Model Cube** and its seven dimensions.

4.2 Design/Methodology/Approach to the Business Model Cube

The methodology used and applied to verify and research on the BM Cube is structured firstly around deductive reasoning. First, a theoretical background of the BM Cube related to business model theory on each dimension of a BM is presented to provide a foundation for commonly accepted and acknowledged dimensions of a BM.

To verify the existence of the dimensions of the BM and the usability of the BM Cube, two business cases out of over 400 are used and presented in this chapter – Vlastuin and HSJD. To "stress test" the generic use of the BM Cube framework, the cases represent two very different test businesses with different BMs. Both cases are chosen to exemplify the concept of the BM Cube in the use of "to-be" and "as-is" BMs. "To-be" BMs are considered under construction – and perhaps lacking one or more of the seven dimensions – and "as-is" BMs are considered to be already operating in the market.

The information and data from the Vlastuin and HSJD cases were gathered through participative action research (Wadsworth 1998) carried out over three years in the EU FP 7 IOT project Neffics (Neffics 2012). Based on these cases, supplemented with other empirical cases and tests, a final definition of the BM Cube concept was formulated in 2011 and is now illustrated in this chapter, along with the detailed test and confirmation of the BM Cube that we also conducted. Appendix 1 shows which businesses the BM Cube has been empirically tested in. The BM Cube has also been functionally tested in cases with different uses on the Neffics BM software platform (Neffics 2012) together with the Dutch ICT provider Cordys (www.Cordys.NL – now www.OpenText.com), the Norwegian Software provider Induct (www.Induct.com) and the Dutch ICT provider VDMbee (www.VDMbee.com). The BM Cube, together with the VDML standard, was proposed in 2011 as an OMG standard (www.OMG.org) and was adopted as an OMG standard in 2013.

4.3 The Seven Dimensions of the Business Model Cube

4.3.1 Value Proposition Dimension

All the business models we checked in our research (Appendix 1) acknowledge that any business offers or proposes values. We define these firstly as the value proposition offered to the customers or users. This can be in the form of products, services and/or processes of services and products. Values are offered by the business as related to the *customer functions* that the business offers to solve for the customer (Abell 1980). Customer values can be: products – a light bubble; services – an installation of a lamp or solutions to the specific lighting of a building; or a value proposition process – a specific process consisting of lamps, installation and lighting through a certain time period delivered in a certain process to the customer. Kotler (1984) supports this argument by expressing that any business delivers or offers values in the form of products and/or services and/or process. (See also Magretta 2002; Osterwalder et al. 2005; Chesbrough 2007; Johnson et al. 2008; Casadesus-Masanell and Ricart 2010; Osterwalder and Pigneur 2010; Teece 2010; Zott et al. 2010; Osterwalder 2011.)

The literature of business process engineering (Davenport 1990; Hammer 1990) increases the value proposition dimension as it argues for a value proposition process. This is further supported by Chan and Mauborgne (2005) talking about a value proposition process before, during and after the carrying out of a certain value proposition exchange. A value proposition process thereby takes in the time aspect of any value proposition exchange and extends the value proposition offer from any business to more than just products and/or services.

4.3.2 Customers and/or User Dimension

All academic works and practitioners we consulted agree that business serves customers and/or users (Chapter 3; see also Appendix 2). “A successful *business* is one that has found a way to create value for its customers – that has found “*a way*” to help customers and/or to get an important job done” (Johnson et al. 2008). “It’s not possible to invent or reinvent a business model without first identifying a clear customer value proposition” (Johnson et al. 2008).

Here, we draw a distinction between customers and users. Customers pay with money – “there is no marked – *Business* – if the customers do not pay” (Kotler 1984), whereas users (von Hippel 2005) do not pay with anything or pay with other values.

Business model theory (see Chapters 2 and 3) until now has only considered the business model related to customers. However, as we will see later, and as von Hippel argued, users can be highly valuable to business by "paying" with other values.

4.3.3 Value Chain Functions (Internal Part) Dimension

Any operating business has functions which are (Porter 1996; Sanchez 1996, 2000) able to "offer" value propositions and serve the customers and/or users with values. Most of the academic frameworks we checked acknowledge this but few are very concrete about which functions are involved and some have not even mentioned these.

A value chain function list could be adapted from Porter's value chain framework (Porter 1985, 1996) including: primary functions – inbound logistics, operation, outbound logistics, marketing and sales, service; and support functions – procurement, human resource management, administration and finance infrastructure, business model innovation. We changed Porter's product and technology development support function to a broader support function, which we call the business model innovation (BMI) function, as we believe that BMI covers Porter's two support functions. The BMI function was not considered by Porter at the time he introduced the value chain model. Porter was, at that time, primarily focusing on products and the activities of the value chain. In Table 4.1 we propose a list of value chain functions (internal part) to be carried out in any BM.

Table 4.1 Value chain functions – primary and secondary function list of any BM

Primary functions	Support functions
Inbound logistics Examples: quality control, receiving raw materials control, supply schedules	**Business model innovation** Examples: innovation on the seven BM dimensions
Operations Examples: manufacturing, packaging, production control, quality control, maintenance	**Administration, finance infrastructures** Examples: legal accounting, financial management
Outbound logistics Examples: finishing goods, order handling, dispatch, delivery invoicing	**Human resource management** Examples: personnel, lay recruitment, training, staff planning
Sales & marketing Examples: customer management, order tracking, promotion, sales analysis, market research	**Procurement** Examples: supplier management, funding, subcontracting, specification
Servicing Examples: warranty, maintenance, education and training upgrades	

Any operating business needs to include some of these functions in some degree – which Porter refers to as activities that are carried out to enable a business to be able to fulfil its purpose, either executed by the business itself or carried out by others. The result of carrying out these functions is value added and/or fewer costs (Porter 1996) which can be proposed as value propositions.

Porter's list was originally described as activities and developed on the background of an operating business. It was not in particular made for "to-be" businesses – entrepreneurs, new or changed businesses, or businesses that were in a "phase of BMI" before BMES introduction or made ready for operation. Our model acknowledges "as-is" activities but we find that it is necessary to include also the functions of a "to-be" BM that is not yet operating and still has activities.

4.3.4 Competences Dimension

Very few BM frameworks (See Chapter 3 and Appendix 2) comment on and address the questions: "How are the activities and functions carried out?" "Who takes care of the value chain functions?" and "By which competences are the value chain functions carried out?" According to Prahalad and Hamel (1990), competences can be divided into four groups – technology, human resources, organizational systems and culture. Technology according to the MIT approach covers product, production and process technologies, human resources cover the employees used in the business and its related business models, organizational systems and culture (Tillich 1951, 1990). The business can choose either to use own competences, network partners' competences and even users/customers competences to carry out the value chain functions.

According to Prahalad and Hamel (1990), any business can have competences but only a few businesses would have core competences. Often it has been said that it is strategically preferable to protect, insource and control core competences within the business itself – and have value chain functions that are not core to the business and business model carried out by other competences, e.g. network partners' competences.

4.3.5 Network Dimension

Håkansson argued that any business is in a network of other businesses and thereby "no business is an island" (Håkansson and Snehota 1990). Any business is a network-based business and these networks could be physical, digital and/or virtual (Goldman et al. 1995; Child and Faulkner 1998; Hamel 2001; Choi 2003; Vervest et al. 2005; Lindgren 2011). Very few of the BM frameworks mention networks; however, historically networks have been more important and visible in the latest 10 years of BM research. Increasing numbers of businesses have chosen to outsource the handling and responsibility of taking care of specific value chain functions. Network partners have in this case been increasingly important in many businesses' business models.

4.3.6 Relation Dimension

Businesses' business models are related through tangible and intangible relations (Provan 1983, Provan et al. 2007, Provan and Kenis 2008; Allee and Schwabe 2011) to other businesses' customers, competences and networks (Håkansson and Snehota 1990; Amidon 2008; Russell 2012). Businesses' BMs are related through strong and weak ties (Granovetter 1973) Businesses' BMs send value propositions to other businesses' BMs through relations and receive value propositions from other businesses' BMs through relations. Relations can be one to one or one to many, visible or invisible to humans or machines (Lindgren 2012).

Tangible and intangible relations are used in the business to deliver values (Allee and Schwabe 2011). Businesses relate their value proposition, users/customers, value chain functions, competences, network and value formulas through relations. Relations are used for creating, capturing, delivering, receiving and consuming values. Value propositions are sent through tangible and intangible relations to users, customers, competences and network. Relations are connected to roles (Allee and Schwabe 2011) played by users, customers, competences and/or network partners.

Very few BM frameworks (Chapter 3 and Appendix 2) include relations. Osterwalder (2011) acknowledges customer relations as the business is related to customers but seems to forget relations to suppliers and other stakeholders in the BM. Only very few (Casadesus-Masanell and Ricart 2010; Allee and Schwabe 2011) go into visualizing and documenting value transfers through relations in the BM. We found in our empirical tests that a BM without relations between the other BM dimensions will never be able to operate and become an "as-is" BM. We also found that relations that are not "connected", independently of whether they are tangible or intangible, cannot transfer values from one BM dimension to another.

4.3.7 Value Formula Dimension

In our empirical tests and research, we found that any business uses some kind of a formula to calculate the value it offers to its own business or any BM in any BMES. Very few BM frameworks (Chapter 3 and Appendix 2) comment on this formula and those who do are quite vague about the formulae.

The value formula is a formula that shows how the value and the cost are calculated by the business (Henry and Haynes 1978; Kotler 1984; Porter 1985; Osterwalder 2002). The result of this calculation is a value formula expressed in money and/or other values. Henry talks about a profit maximization formula, Kotler talks about several pricing models, Porter discusses different competitive pricing formulae and Osterwalder (2011) expresses this in his BM framework as revenue and cost structure. Very few academics dealing with BM deal with how the business calculates the value they want to get out of the BM.

4.3.8 Business and Business Models

The seven dimensions mentioned above are equivalent to the overall model – the BM Cube – that we propose describes how any business model is constructed. The seven dimensions should be considered by any business that is interested in running its BM operations well.

However, we found in our research that there is a difference between the way businesses want to run their operations – seven visionary dimensions of a business – and how a business really runs its operations. By mapping empirical data from our business case studies to the seven dimensions, we found that most businesses have more than one business model. In other words, the businesses described via the seven dimensions are different to how these businesses actually run their BMs. Some of these BMs were close to their original description of the seven dimensions but others were different.

This attracted our attention to the fact that businesses could potentially have more business models and that there could exist a level beneath the business's seven overall dimensions. We therefore address the importance of investigation of these business models and draw a distinction between a visionary model of a business and the models of business that are actually carried out ("as-is") and that are intended to be carried out ("to-be") in the business.

Most academics working with BMs have until now used the term "BM" at the business level and at the visionary level. Further, they use it to cover just one BM for any business, as seen in Table 4.2.

This observation, together with inspiration from Abell's and Hamel's original definitions and framework of "the core business" (Abell 1980) and "the core competence" (Hamel and Prahalad 1994) made us adapt the definition of "the core business model" as the BM model at a business level and business visionary level, which states how businesses related to the seven dimensions may wish to run their businesses. In this context we found on behalf of our research it was necessary to increase Abell's dimensions from

Table 4.2 Business model definition focal points

Authors	BM as framework	BM at business level	BM at business model level
Abell (1980)		X	
Timmers (1998)	X		
Venkatraman and Henderson (1998)	X		
Selz (1999)		X?	
Stewart and Zhao (2000)		X	
Linder and Cantrell (2000)		X	
Hamel (2000)	X		
Petrovic et al. (2001)		X?	
Weill and Vitale (2001)		X	
Magretta (2002)		X	
Amit and Zott (2001)	X		
Markides and Charitou (2004)		X	(x)
Malone et al. (2006)		X	
Chesbrough (2007)	X	X	
Skarzynski and Gibson (2008)		X	
Johnson et al. (2008)	X		
Casadesus-Mansanell and Ricart (2010)		X	(x)
Johnson (2010)		X	
Osterwalder and Pigneur (2010)		X	
Teece (2010)	X		
Zott et al. (2010, 2011)		X	
Fielt (2011)		X	
Lindgren and Rasmussen (2013)	X	X	X
Gassmann (2014)	X	X	

Note: Where X? appears, we had difficulties in placing the X, precisely due to a kind of fuzziness about what the authors really mean and focus on in their frameworks. Therefore, the placement of X? is our indication of where they should be or we, based on their descriptions, think they are placed.

three to seven dimensions and added some components to his dimensions that were lacking.

The core business model's seven dimensions refer hereafter to:

> How a business wants to construct and intends to operate its "main" and "essential" business related to the seven business model dimensions – value proposition, user and/or customers, value chain (internal functions), competence, network, relations and value formula.

The business model (BM) refers to:

> How a certain BM in the business is constructed and actually operates – "as-is" BM – or is intended to be constructed – "to-be" BM – related to the seven dimensions: value proposition, user and/or customers, value chain (internal functions), competence, network, relations and value formula.

In other words, businesses most often have both "as-is" BMs and "to-be" BMs, which we will comment on in Chapter 5, which discusses the multi business model approach.

4.4 The BM Component Level

Each BM dimension can be divided into "smaller parts", which we call components (Appendix 7). We will now exemplify the BM dimensions by explaining how each dimension in any BM can be and most often are constructed differently on the component level. We will show how they can be characterized on a BM component level. The level of detail of each dimension is up to the individual business to decide. Business can "dive" as deep in detail as it wishes; however, our research and theory show that examining components can give meaning and it is highly valuable to the business to go into detail. Businesses must be able to get value out of the details – the components – otherwise they will miss the overview and motivation of mapping their BMs.

4.4.1 The Value Proposition Dimension Component Level – "What Value Propositions Do the BM Provide?" (VP)

The definition of value (Alderson 1957; Drucker 1973; Anderson 1982; Albrecht 1992; Woodruff 1997; Anderson and Narus 1999; Doyle 2000; Lindgreen and Wynstra 2005; Wouters et al. 2005; Chan and Mauborgne 2005; Osterwalder et al. 2005) is manifold and its development since the 1950s during the "era of innovation" has been covered intensely in academia.

Value is key in understanding the value of a product, service, process and relationship offered. However, value proposition varies as it relates to different customers, because just as customers are different they are also satisfied by different values, whether it is from products, services, a relationship or a value fulfilment delivered in a process by products and services (Lindgren 2011). "Managers today continuously ask themselves: How can we understand customer's value and how can we deliver 'real' value to customers in a cost efficient and profitable way?" (Johnson et al. 2008).

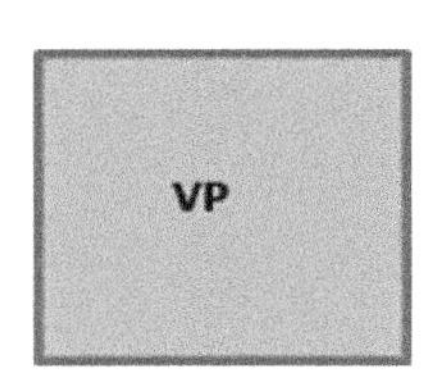

Figure 4.3 The value proposition dimension.

The customer's value equation is often very complex to understand in detail because it is not static but dynamic over time (Lindgreen and Wynstra 2005). Therefore, value proposition has to be understood from:

- the perspective of both the business and the customer and/or user the value is delivered to
- the context the value is delivered in
- the time the value is delivered
- the place the value is delivered
- the relations the value is delivered through

Value can be said to be closely connected to the concept of "total value and cost to the customer" (Wouters et al. 2005). In this case, staying at the point of entry to a trade or a value proposition process is strongly related to the customer's total perceived value and total perceived cost related to the products, services or process. This is why it is incredibly difficult as a business to measure and read the values and cost of a customer, and to decide the degree of attractiveness of a value – or whether a value is judged high or low related to a trade or a process. In this chapter, we focus on what the business – or business model – believes it offers related to value: the business viewpoint (Lindgren 2011). However, we acknowledge that there are also other views of a value proposition.

The solution to classifying value propositions taken by many businesses is to offer different value propositions to different customers, which argues that a value proposition offered by a business is often different for each **customer, context, time** and **place**.

Payne and Holt (1999) outline four types of values.

1. **Use values** – the properties and qualities which accomplish a use, work or service for the customer
2. **Esteem value** – the properties, features or attractiveness which cause the customer to want to own the product and service
3. **Cost value** – the sum of labour, materials and various other costs required to produce a product or service for the customer
4. **Exchange value** – the properties or qualities of a product or service, which enable exchanging it for something else that the customer wants

We found that this list of types of values had to be complemented by an overall dimension of work time vs. life time (Fogh Kirkeby 2003). Time as the factor that defines customers' personal or business values of the, for example, trade or process is related to an overall lifetime value and describes the sum of actions taken in order to find work life-fulfilling and transcend oneself, a value often seen as the driver of projects, art etc. (Tillich 1951; Austin and Devlin 2003; Sandberg 2007).

Value also has to be measured **before, during** and **after** value exchange has taken place (Kim and Mauborgne 2005). This means that a customer could trade or collaborate on the value from a product and service that comes out of the trade (Kotler 1984; Ziethaml 1988; Doyle 2000) but also from the value of the relationship (Reichheld 1993; Lindgreen and Wynstra 2005). The creation, capturing, delivering, receiving and consumption of value through a relationship (Brodie, Brookes and Coviello 2000; Lindgreen, 2001; Coviello et al. 2002; Lindgreen, Antioco and Beverland 2003; Lindgren 2012) is the value equation of an inter-organizational collaboration project – a network-based BM. This is one important value and also an attraction factor, which could be in this case an innovation of a "to-be" business model. The value of this can be something other than money, e.g. learning. There is a list of non-monetary values in Appendix 3.

This is in line with research claiming that the value of the relationship, activity links, resource ties, and actor's bonds (Håkansson, 1982; Axelsson and Easton 1992; Håkansson and Snehota 1995; Ford 2001; Ford et al. 2002; Ford et al. 2003) can be even more important than the value of the product or service. The value of the relationship is both an input and an output of the business model innovation process, which supports the argument that value is not static but dynamic.

As values are created, captured, delivered, received and consumed in a value process; they are continuously undergoing change throughout the business model innovation process or the lifetime of values. Values of relationship can be related directly (e.g. profit, volume and safeguard functions) and also indirectly (e.g. innovation, market, scout and access functions). The value functions (Walter et al. 2001) can further be of a low- and/or high-performing character (Lindgreen and Wynstra 2005) which is often up to the customer's judgement and to influence the degree of this value. Kim and Maubourgne express this in their strategic value map (see Appendix 4). However, their value map is just seen from the business viewpoint and not from the customers' or other viewpoints (e.g. network, value chain function, competence, relation and process viewpoint) (Lindgren 2011).

The value of a customer should also be understood as perceived value – benefits and cost (Woodruff 1997; Walter et al. 2001; Lindgren and Dreisler 2002), which means that the real value of a product, service and/or a process can in some cases be neglected in favour of a higher or lower perceived value of a product, service or process. Furthermore, perceived value should not just be related only to the individual customer but alo to other individuals as customers, users (von Hippel 2005), competences (technology, humans, organizational systems and culture) and network (suppliers or other networks) in the business model interpretation of the product, service and/or process (Blois 2004). Therefore, it is the user's, customer's, competence's or network's interpretation of "value" that is important and not just what the business and its stakeholders (investors, the market, the business, the innovation leader) "think" ought to be or are the values – that is the real value proposition of the BM. In Part 2 we will comment on these different views of value – refering to the BM panorama view.

It is therefore important when analysing and understanding a product, service and/or process value, to analyse all stakeholders and both values and perceived values. Furthermore, it is important to analyse values and perceived values over time, during the trade or inter-organizational collaborative process, as both values and perceived values are dynamic and will therefore by definition always change throughout the entire value process and thereby over time. Today no business model framework has managed to cover and capture value change over time.

Values can be tangible and/or intangible. "Tangible" describes something you can see, touch or feel and others can get a full view of these components. Intangible values you cannot see, touch or feel physically. Sometimes, however, you clearly understand that the intangible values exist and have an impact – maybe even more than the tangible values.

We make a distinction between tangible and intangible values and associated value objects. Tangible value objects have often a direct financial value, underpinned by an accepted financial marketplace for realizing the value.

A view of tangible and intangible values is inspired by Verna Allee's framework (Allee 2008), which defines tangible values as deliverables to include anything that is contracted, mandated or expected by the recipient as part of the delivery of a product, service and/or process that directly generates revenue. Intangible value objects, as proposed by Allee, could be considered in three main groups:

- Intangibles where a financial market may be established but where the stability and absolute nature of the value may be questionable (such as intellectual property).
- Intangibles where a measure is established with a wide acceptance of the measurement approach (such as a carbon footprint).
- Intangibles where only a specific context is applicable with values very much related to that context.

Li et al. draw a comparison between tangibles and intangibles in relation to markets and contexts (Li et al. 2012). This enables us to include the operation of social businesses/exchanges within this definition of tangibles and intangibles.

In summary, any business model may offer a value proposition, which can be offered as tangible and/or intangible value. Value proposition can be products, services and/or processes of product and services. Value propositions can be values of relations.

4.4.2 Customers and Users Dimension Component Level – "Who Does the BM Serve?" (CU)

Any business model that we researched talks about business models having customers. However, we found that many BMs do not have customers that pay for BMs' value proposition, but are constructed around users, which provide the foundation for other BMs with customers. Facebook, Skype, LinkedIn, Twitter and Google are good examples of such business models. Ryanair, Uber, Airbnb are examples of business models where the customers do not pay the real costs of "production" of the BM's value proposition. How can this be? And – how can and should we understand this from a sustainable business and "going concern" perspective?

Our research showed that BMs built upon users, when growing big in numbers of users, can attract and activate customers willing to buy – or pay for – value propositions in other BMs. Either users start to pay for better performance, advanced use, deeper content, for example, or other customers buy, for instance, promotion, because there are so many users in the BM. In these cases, the customers pay for other or different value propositions – or even a different BM – compared to the users. Stock buyers of the Facebook business could be an example. The customers, however, can play a double role also – at the same time being users of the value offering in the user- based BM. Stock buyers of the Facebook business are probably often also Facebook users. Thereby customers can play different roles in a BM and in different BMs. This complicates the "picture" of business models.

This is one of the arguments for the existence of more BMs (Lindgren 2012, Lindgren and Rasmussen 2013). In all businesses where our research

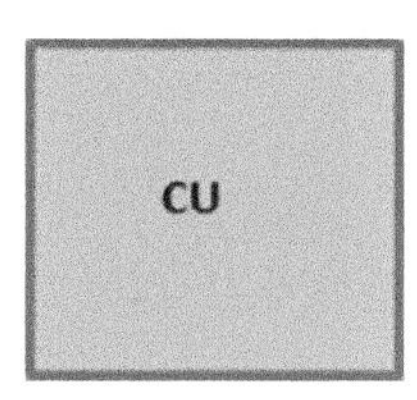

Figure 4.4 The customers and users dimension.

was carried out we found more BMs and that BMs were often interrelated and added value and influenced each other.

We therefore propose to distinguish between users and customers by defining users as not paying for the value proposition (Kotler 1984; Von Hippel 2005) while customers pay for the value proposition (Kotler 1984).

Users can, however, "pay" with other values, other value transfers and thereby contribute to development of very important values for other business models. These values could be learning for future BMI, development of critical user mass that would be attractive for other BMs, and change of general market context and direction. Needless to say there can be many other valuable contributions from user-based BMs to customer-based BMs (Appendix 3).

4.4.3 Value Chain Functions (Internal) Dimension Component Level – "What Value Chain Functions Do the BM Have?" (VC)

Any business model must carry out certain activities to produce the value proposition for the users and/or customers. A list of these activities was proposed by Michael Porter in his value chain framework (Porter 1985). Porter called these "functions" and proposed some primary functions and some secondary functions to be carried out by a value chain. A value chain was proposed by Porter to include one or all of these functions; however, if some functions were missing and not carried out, our research showed that this can stop the BM's operations or that the BM will never come to operate in the business and the business model ecosystem (BMES).

Porter's value chain framework was related to an operating BM. However, when businesses start to create a "to-be" BM there are really no active functions, just wishes and expectations of value chain functions the BM should carry out. Further, when we observe an operating business at a certain moment – in this case, we freeze the picture of a specific BM – we do not see "running" activities but just functions that are carried out (Appendix 5). Value chain functions in our BM framework represent the value chain functions that

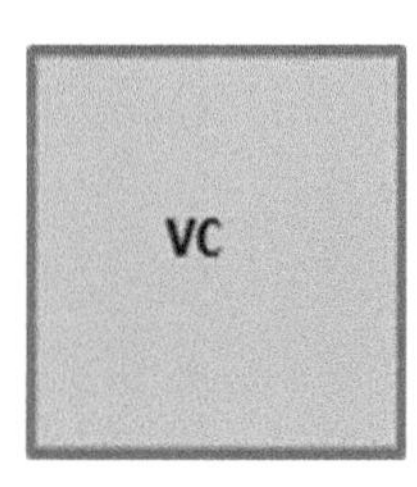

Figure 4.5 The value chain function dimension.

have to be carried out or are being carried out within the BM. We acknowledge that there are value chain functions outside the BM but in this chapter we only focus on the internal value chain functions of the BM.

4.4.4 Competence Dimension Component Level – "What are the BM's Competences?" (C)

Any business models rely on and use competences, either from the focal business, from network partners or even from customers and users to carry out the value chain functions that create, capture, deliver, receive and consume the value propositions.

As we have seen, according to Prahalad and Hamel (1990) competences can be divided to four main categories: technologies, human resources, organizational systems and culture.

Technologies, according to (Sanchez 1996, 2000, 2001), are divided into:

1. Product and service technologies
2. Production technologies – both "product- and service-production technologies"
3. Process technologies – to run and steer the production technologies so that the product and service technologies can be created, captured, delivered, received and consumed.

Each BM has a specific mix, integration and use of product and service technologies, production technologies and process technologies. Sometimes the mix, integration and use of technologies is so unique that the competence can be classified as a core competence (Prahalad and Hamel 1990).

Human resources are the people – either white collar or blue collar (Peters 1997) – that the BM can use to carry out the value chain functions. The human resource, its mix and its use can also be so unique that human resource too can be rendered as a core competence.

Organizational systems are the systems that the business models use to organize the use of technologies and human resources to carry out the value

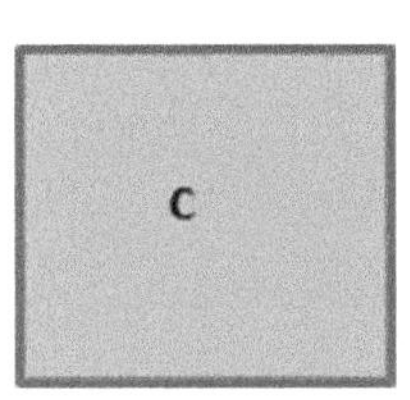

Figure 4.6 The competence dimension.

chain functions. The organizational system can also be so unique that it is a core competence.

Culture is the "soft" part of the competence dimension. We claim that any BM has a specific culture. The culture can be adapted one to one from the business or other BMs but can also be incrementally, even radically, different from these. Most users, customers, employees and networks "feel" the culture and the difference in culture when entering or dealing with a business – either it is physical, digital or virtual.

4.4.5 Network – "What is the BM's Network?" (N)

In our research we found that any business model is network based. No BM is a lonely island – at least not for a very long time. Why? Because if a BM does not receive value from outside it will slowly shrink and vanish. If it does not offer a value proposition of any kind it will not be able to receive value in a long-term perspective. The BM network thereby becomes vital to any BM – a BM is its network.

Networks can be physical networks (Håkonsson and Snehota 1990), digital networks (Choi 2003) and/or virtual networks (Goldman et al. 1995; Vervest et al. 2005) that the BMs use.

4.4.6 Relations Dimension Component Level – "What are the BM's Relations?" (R)

Any business model relies on relations. Relations in our terminology enable BMs to transfer value from one BM dimension to another. Relations enable

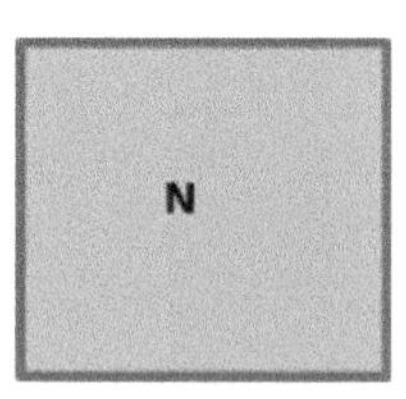

Figure 4.7 The network dimension.

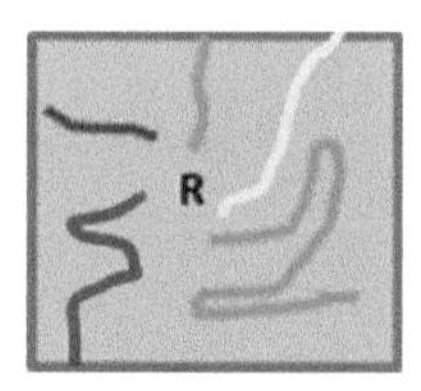

Figure 4.8 The relation dimension.

BMs to create, capture, deliver, receive and consume values. Relations are like the "arteries", "veins" and "nerves" in the "body". Relations can have forms as tangible and intangible relations.

In our initial research, we found four sets of relations that were of importance to BMs (as shown as examples in Appendix 5) and that should be attended to by business managers. See Figure 4.9.

1. The **"inside BM inside business"** area relations – business model relations transferring values and securing communications inside the BM.
2. The **"inside business outside BM"** area refers to relations between different BMs inside the business.
3. The **"inside BM outside business"** refers to relations between BMs outside of the business.
4. The **"outside BM outside business"** refers to relations and relation areas where the BM and business do not share a relation.

We will elaborate more on the relations axiom in Chapter 7.

Value and values of a BM can be seen in a broader perspective as each partner's BM's relation to users, customers, competences and networks in the inter-organizational network of relations to "as-is" and "to-be" BMs. Why? Because value and cost are strongly interrelated with relationships (Blois 2004), and attributes related to the relationship between the partners' BMs in, for example, a simple trade "as-is" BM or a BM innovation project "to-be"

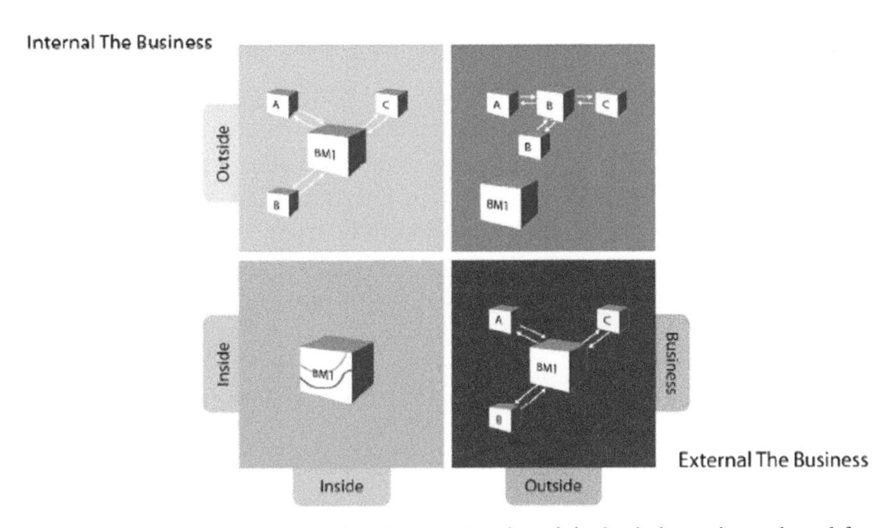

Figure 4.9 The relations areas related to a BM – the original relation axiom adapted from Lindgren and Rasmussen 2013.

BM where goods and services are not necessarily defined. Needless to say, these relations also influence each other and are interrelated. However, this is not studied much in BM literature.

As was seen earlier, value proposition is not only related to products, services and processes but is also strongly connected to relations and thereby a result of the relation between BMs in either a trade or a BMI project. Value equation can be related to irrespective of whether the BMs are related or not. In this chapter, we only cover the internal relations – the "in in" relations – in a BM.

Relations, activity links, resource ties and actor's bonds (Håkansson 1982; Axelsson and Easton 1992; Håkansson and Snehota 1995; Day 2000; Ford 2001; Ford et al., 2002; Ford et al., 2003) are all tools used to describe and map relations.

The creation, capturing, delivering, receiving and consumption of value is enabled through relations (Brodie, Brookes and Coviello 2000; Lindgren 2001; Danaher and Johnston, 2002; Lindgreen, Antioco and Beverland, 2003, Lindgren 2012). Relations connect the different BM dimensions' components and enable the creation, capturing, delivering, receiving and consumption process of value. However, if any BM is not able or "willing" to send and receive the value through the relations, then the relations have no value and no task for the BM. Therefore it is very important for managers of businesses and managers and participants of BMI projects to focus on the relations of BMs.

4.4.7 Value Formula Dimension Component Level – "What are the BM's Value Formulae?" (VF)

In our research we found both theoretically and empirically that any business model will have one or more value formulae. The value formula can be expressed in either a monetary and/or in a non-monetary way. The term "profit formula" as a dimension in a BM that we found through our research is too narrow a term for BMs and – we propose – has to be changed to a dimension called "the value formula dimension" to cover all types of BMs. We found that profit formula is too narrow a term to express the formula by which the value

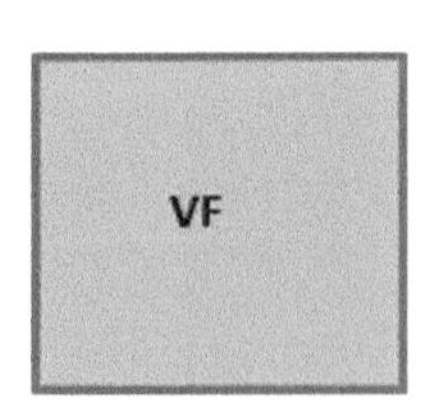

Figure 4.10 The value formula dimension.

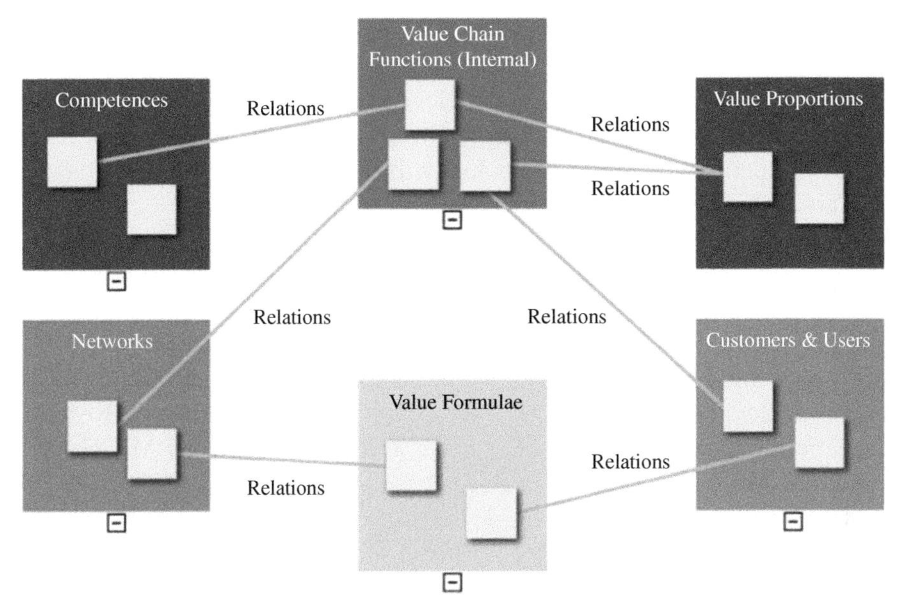

Figure 4.11 The seven dimensions of the BM Cube shown in a 2D presentation.

of a BM is calculated because our research showed that many businesses and BMs are not focused, or are not exclusively focused, on profit but instead on other values – value formulae. They "calculate" on other value formulae and to get a full understanding of why business models exist and are innovated it is necessary to include other values. We therefore propose profit formula as one of many value formulae that can be the "calculated" output of a BM. However, we claim that any BM has one or more calculated value formulae – monetary and/or non-monetary. A BM can have more than one value formulae.

Having verified academically that the seven dimensions of the BM exist enables us to complete the concept of the BM Cube. In a 2D picture and with the seven dimensions spread out flat it could look like the sketch in Figure 4.11.

However, we discovered that the seven dimensions form a BM Cube with the "in in" relations inside the Cube as shown in a sketch model in Figure 4.12.

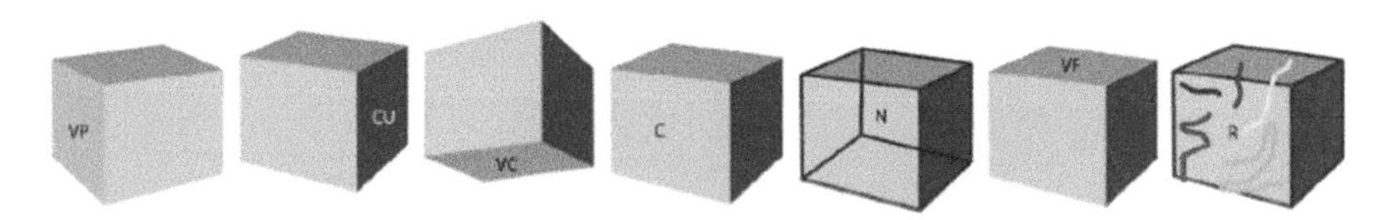

Figure 4.12 The seven dimensions of the BM Cube presentation.

The 2D version is very helpful when working on a BM dimension level but the 3D version can be even more helpful when working on a BM, BM portfolio, business and BM ecosystem level. Both presentations are helpful when working on BMI but a strong digitization of the BM – as we will comment on later in Part 2 of the book – will be extremely helpful in the future. This will enable us to "dig deep" in any business model.

4.5 Summary

Summing up, we propose that any BM Cube consists of seven dimensions – six sides and the BM relations inside the BM Cube that bind all other dimensions and components together and enable creation, capturing, delivering, receiving and consumption of the values that lie outside the BM Cube and bind the BM together with other BMs. We illustrate the BM Cube in Figure 4.13.

Any BMs can be defined as related to the generic BM concept consisting of seven generic dimensions. Each of the seven dimensions addresses some core

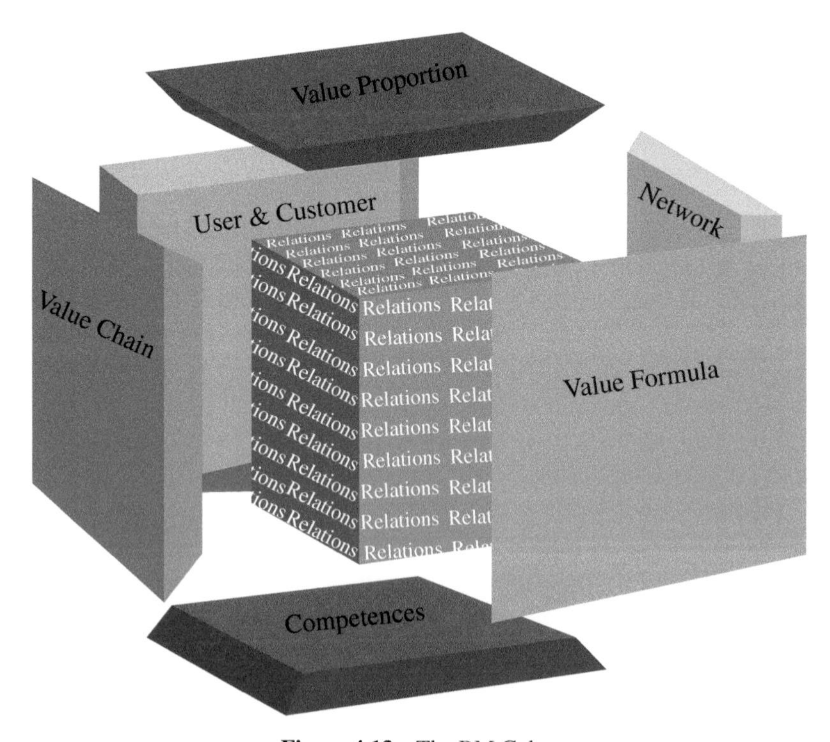

Figure 4.13 The BM Cube.

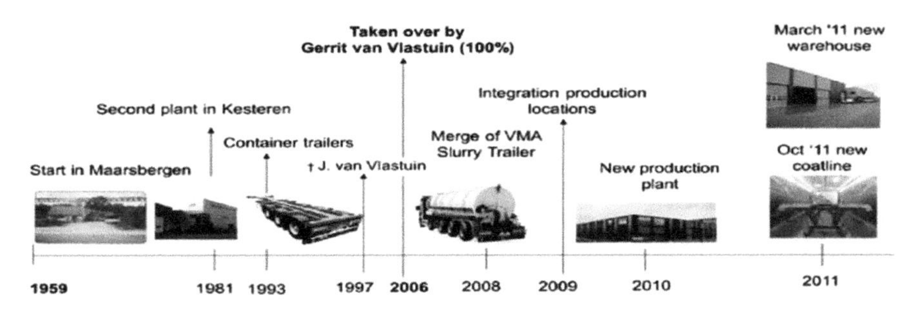

Figure 4.14 Vlastuin's business evolution.

questions in relation to each individual BM's dimensions characteristics and logic (see Table 2.1 for these dimensions and questions). Each BM dimension can be split into small BM dimension components.

With the above mentioned it is now possible to draw up the first part of the vertical butterfly model (Rasmussen, Lindgren and Saghaug 2014; Lindgren 2016b) as seen in Figure 4.14. These levels we will comment on more in Chapter 5.

4.6 Business Cases

In order to approach the combination of business and BMs and to define, visualize and document the BM Cube, two case studies are presented as examples as a follow-up to Chapters 1 to 4.

The first case is based on the Dutch business Vlastuin which is implementing several new "to-be" BMs in order to reinforce its business and already has several BMs operating as "as-is" BMs in order to sustain its business. The second case is concerned with an already functioning hospital in Spain, HSJD, which introduced a whole range of "to-be" BMs in relation to the hospital's business.

Here we give a very brief description of the two business cases. Further details can be found in Appendices 5 and 6.

Vlastuin (Appendix 5)

Vlastuin, located in Netherlands, started its operations in 1959. Vlastuin employs around 150 people and had a turnover of 27 million euros in 2011. During its more than 50 years, Vlastuin has added more BMs to its business and thereby slowly increased its core business. It started off by installing and servicing furnaces and boilers, gradually moved to manufacturing and later on added assembling of cranes and parts to the business. A graphical representation of Vlastuin business evolution can be seen in Figure 4.14.

In Appendix 5, a detailed description and analysis of the case is presented.

HSJD Hospital (Appendix 6)

Hospital Sant Joan De Dieu (HSJD) belongs to the Hospital Order of Saint John of God and is a private, non-profit hospital. The order is represented in more than 50 countries and has almost 300 healthcare centres worldwide. HSJD is located in Barcelona, Spain, and is a children's and maternity care centre. It is a university hospital connected to the University of Barcelona and is also associated with the Hospital Clinic of Barcelona, which helps the hospital to provide high-level technological and patient care. HSJD is 95 per cent financed by the Catalonian public system and the remaining 5 per cent comes from private investments. The primary goal of HSJD is to encourage and educate people to follow a healthy lifestyle with good nutrition, proper sleep, hygiene and exercise.

In Appendix 6, a detailed description and analysis of the case is presented.

5

The Multi Business Model Approach

Peter Lindgren

Abstract

Today, most academics and practitioners consider the business model (BM) as measurable, objective and one of a kind. Although there are many different definitions (Taran 2011) and types of BMs (e.g. open and closed (Chesbrough 2007; Lindgren 2011), free (Anderson and Narus 1999) and internet-based (Zott and Amit 2002), most define "business model" on a business level and on a core business level (Abell 1980). In this chapter we propose that there is a need for a distinction between levels of business model focus: the business level – the core business model or overall business model – and the business models existing under the "umbrella" of the core business model. This is to prevent fuzziness and support discussion and help further development of the BM theory and the knowledge of the BM community.

5.1 Introduction

In our research we found that most businesses do not stick strictly to their core business and how they want or have planned their "as-is" business model (BM) to look like and be. They have, in fact, often a variety and a mix of BMs – both "as-is" and "to-be" BMs with different value propositions, users and customers, value chains with different functions, competences, networks, relations and value formulas. One set of seven dimensions does not, therefore, fit all business models, markets, industries and worlds (Lindgren 2011) – the business model ecosystem (BMES). This mix of the seven dimensions – which we classify as different business models, whenever they are different or changed – exists and coexists within the core business. Each individual BM is not – as we said before – necessarily aligned strictly to the core business model and the seven dimensions of the core business model. All of them have their own specific seven dimensions and all of them show different combinations of the seven dimensions.

We argue, therefore, that a business's different business models cannot be explained by just one business model – "the core business model" – but would preferably be better explained by more and, in fact, by different business models. However, each BM still can and should be explained with the seven generic BM dimensions, but each with their different characteristics on one or more dimensions. In our research, we only found Casadesus-Masanell and Ricart (2010) (see Figure 5.1) and to some extent Markides (2004) who theoretically indicate our findings – and the existence of more BMs in a business. However, we see these as possible strategies for BMs or a BM plan but as different BMs that could coexist at the same time in the business or could be co-innovated and operated in the business. As we see it, one of the reason why this track is not followed is that previous strategy lessons and BM theory did not and cannot cope with more strategies and more BMs. As we learned it several years ago from Prahalad and Hamel (1990) – "stick to your core business" is the best strategy. However, as we argue later, the one does not exclude the other – but we have to change our mindset and acknowledge that BMI and strategizing BMs are complex and will be more complex in the future.

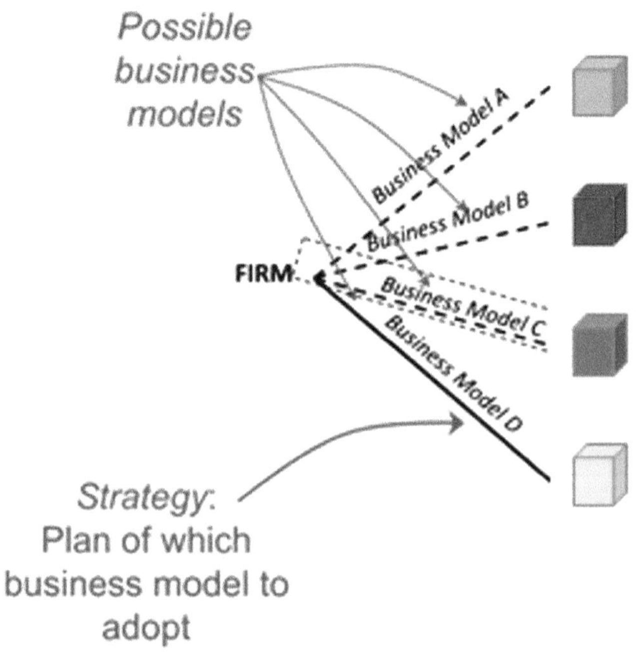

Figure 5.1 The multi business model approach indicated by Casadesus-Masanell and Ricart 2010 related to different operating business models ("as-is" BMs).

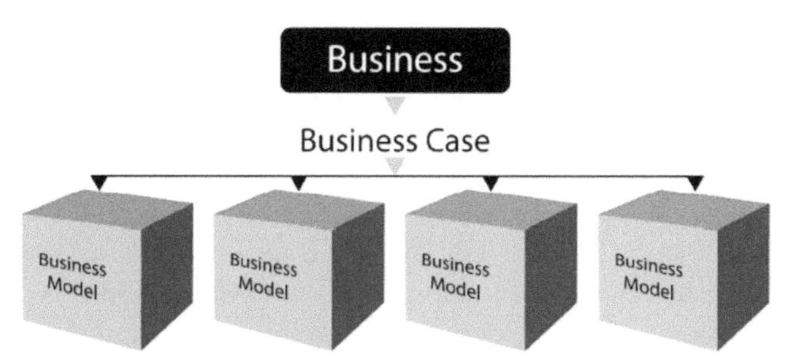

Figure 5.2 The multi business model approach related to different operating business cases and business models ("as-is" BMs).

Most academics only discuss "as-is" BMs: "What are your BMs as indicated ([x] or [x])?" If we illustrate this in a multi business model approach, we would get a picture of more operating BMs – or "as-is" BMs, as seen in Figure 5.2.

In other words we find that businesses most often have a multitude of "as-is" BMs (BM Cubes with unbroken lines) but we also find that they have a multitude of "to-be" BMs (BM Cubes with dotted lines) they are working on – innovating. We believe these BMs have to be seen together – as indicated in Figure 5.3.

In our research we found that "to-be" BMs often influence business operation and performance very much and vice versa. It is therefore necessary to get the full picture of the business and to "download" and "see" both the "to-be" BMs and the "as-is" BMs.

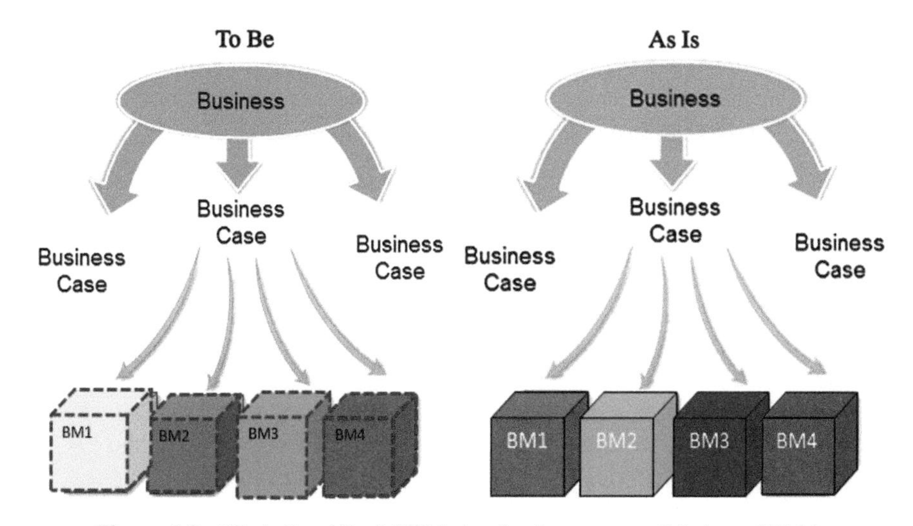

Figure 5.3 "To-be" and "as-is" BMs in a business source (Lindgren 2016a).

As a consequence, we propose that a business can be said to have one or more operating BMs ("as-is" BMs) related to different business cases – the multi business model approach (Lindgren 2011, Lindgren 2016a) – which are more, less or not aligned with the core business model. Further, we propose that a business also can be said to have one or more BMs that are under construction ("to-be" BMs). The "to-be" BMs (dotted-line BM Cubes in Figure 5.3) are in the business model innovation (BMI) phase. The "as-is" BMs are in operation and are fully developed and introduced to their business model ecosystem.

The multi business model approach is, as we will see later in Part 2 of the book quite useful in the understanding of: "What is the business actually doing?" And also useful when we are analysing: "What are competitors actually doing?" and "What are our customers and network partners doing?"

5.2 The Bee Board

In 2014 we developed a very simple tool or board that we called "The Bee Board". We developed it through more iteration together with several SME businesses and entrepreneur businesses. We found it could help businesses to visualize the business BM's – both "as-is" and "to-be" BMs and provide them an overview of their BMs. In Figure 5.4 we show a sketch of such a mapping.

The general idea behind the Bee Board is that when a business model is placed over the horizontal line – "green area" – it generates a positive earning (turnover – cost = profit). When one business model is placed below the horizontal line – "red area" – the business model makes a negative earning – a loss. However, the same board can also be used if measurement is related to other positive or negative values than money. The only issue is for those mapping on the Bee Board to agree on the scaling values on the X and Y axis.

The phases – idea, concept, prototype, implementation, introduction, growth, maturity, decline – follow and are adapted by the general concepts and models from theory about innovation (Cooper 1993, 2005) and the development of a product or a service (Kotler 1984). We even put in some indication lines from theory – but that is purely for inspiration and theoretical trend line indication.

If we go back to the Bee Board and continue to use the money as a measurement guideline then when a BM has a negative earning (loss) on the bottom line it is placed under the horizontal line in the the red field (C and D) and when a BM has a positive earning (profit) on the bottom line it is placed above the horizontal line in the green field (A and B).

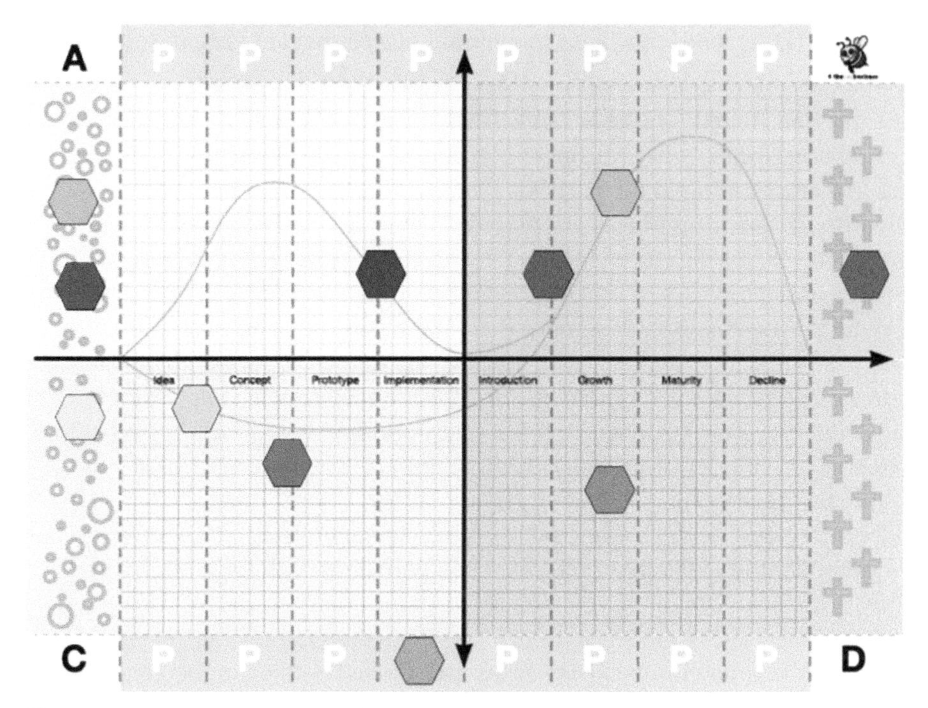

Figure 5.4 The Bee Board – a business model mapping tool for mapping "to-be" and "as-is" BMs ©The BeeBusiness.

When one of the business models is placed to the left of the vertical line, this business model is under development, construction or, as we say, in the BMI phase. When a business model crosses the vertical line, it is fully developed and have entered the BMES. The business has invoiced a customer and received its payment for the value proposition – money. Thereby the full market circle (Kotler 1984) has been achieved from value proposition creation, capturing, delivering, receiving and consumption both for the business and its customer(s).

The Bee Board was originally divided into four fields.

The light green field (A) indicates that this is where business models are placed that are under development and are making a profit. This could be funded BMI projects or where the customer pays in advance for the BM – e.g. Crowdfunding.

The dark green field (B) is where business models are placed that make a profit – when they have been put into the BMES – operating BM's.

Pink field (C) is where the business models are placed that are under development, construction and BMI but are costing business resources, time and money.

Red field (D) is where business models are placed that are fully developed and have already entered the market but are making a loss.

At the **Bee Board's bubble field,** BMs' ideas for new BMs are placed. The gravestones show where business can place their "dead" BMs – BMs that are no longer operating in the BMES.

Bee Board parking places are placed all over so that BMs that are waiting for some outside development – e.g. technical, regulative or business-wise development can be "parked" until the BMI or further operation can take place.

We tested the Bee Board in more than 400 businesses and over 250 education and workshop sessions. We discovered numerous possibilities and variations that the Bee Board can be used for with advantage. In the process we adjusted the Bee Board through several iterations based on our empirical data and feedback during workshops, seminars, educational sessions and try-out in businesses.

Picture 5.1 shows an example where a management group from a Danish valve manufacturer is mapping four business portfolios with their "as-is" and "to-be" BMs.

The multi business model approach can be elaborated further on, which we show in the following business cases and discuss further in the chapters that follow.

5.3 The BM Portfolio Approach

As can be seen in Picture 5.1, the management group of EV Metalværk are actually working with four Bee Boards, where they are mapping their business

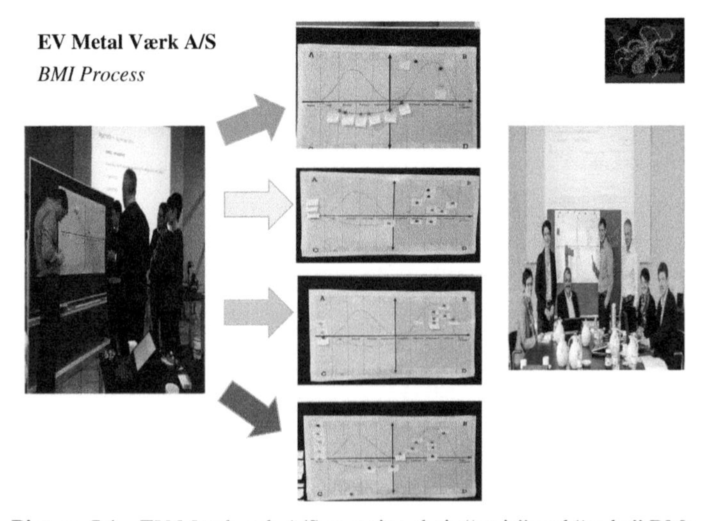

Picture 5.1 EV Metalværk A/S mapping their "as-is" and "to-be" BMs.

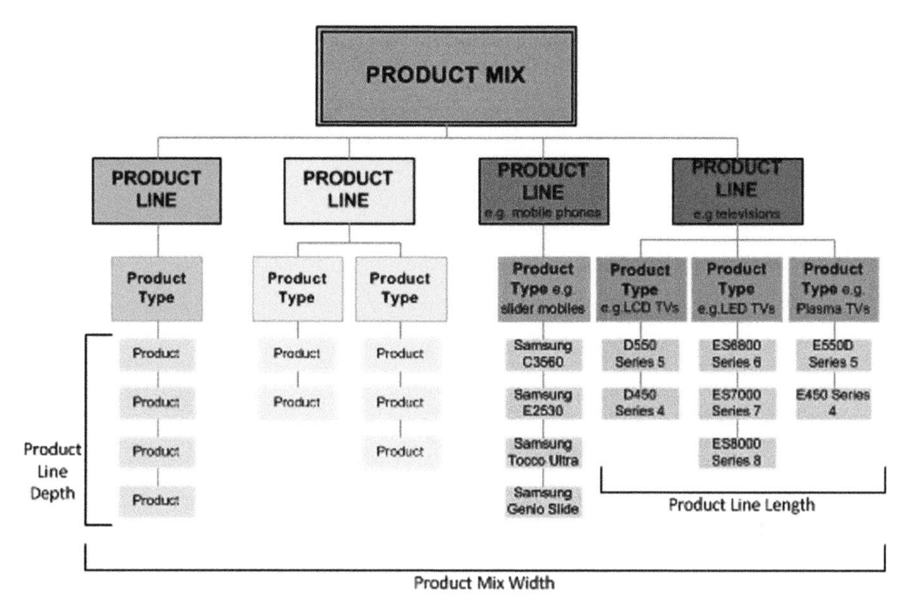

Figure 5.5 Product line and depth by Kotler 1984.

models. In our research we found in many businesses that some BMs were very egalitarian, worked together as in a group or were innovated on the same value proposition, user or customer group, value chain function, competence or network "platform".

We therefore very early in our research discovered that some BMs can together form a group of BMs – what we call a portfolio(s) of BMs in the business (Lindgren 2011). The BM portfolio approach is very much inspired by Kotler's product line and product depth approach (Figure 5.5).

BMs that are interrelated we believe can be grouped and can be treated strategically and tactically as a group. Each BM's portfolio group can be innovated as one group with advantage by the business.

If these BMs form a group of BMs that have similarities due to, for example, the same type of value proposition or customer focus, use of the same value chain or use of the same network it is possible to work with them in the business as a group of BMs. Often we found in our research that the BM portfolios' BMs are interdependent and work as a group independently, and are in the business treated as such. In EV Metalværk A/S they work with four valve types – high pressure, medium pressure, low pressure and hydro ball valve groups.

As we have seen, some BMs attract users who attract customers to other BMs in the BM portfolio. An example of this is shown from the case study

Table 5.1 An example of a portfolio of BMs in the KB use case (Lindgren 2012)

KB BM Case

Business
Business Case
Business Models

	KB Lottery Business Model	KB Cancer Disease Business Model	KB Cancer Information Business Model	KB Cancer Patient Support Business Model
Core Building Block	Building Block	Building Block	Building Block	Building Block
Value proposition/s (Products, Services and Processes) that the company offers (Physical, Digital, Virtual)	Lottery, possibility to win money, things and services	Research aimed at Fighting Cancer disease	Development of Information material aimed discovering and Fighting Cancer disease at an early stage, preventative	Developing and running Cancer Patient Support
Users and Customer/s (Target users, Customers, Market segments that the company Serves geographies, Physical, Digital, Virtual)	Donator who wants to support research in cancer, support of cancer patients, information about cancer	Donator who wants to support fight of Cancer Disease	Donator who wants to support development of information about Cancer Disease	Donator who wants to support development and running Cancer Patient Support
Value Chain (Internal) configuration. (Physical, Digital, Virtual)	KB value chain functions necessary to run the KB lottery	KB value chain functions necessary to handle funding support for cancer research	KB value chain functions necessary to handle information about cancer research, cancer patient support activities, cancer discovery and protection	KB value chain functions necessary to handle Cancer Patient Support
Competencies (Assets, Processes and Activities) that translate company's inputs into value for customers (Outputs). (Physical, Digital, Virtual)	KB technology, HR, organisational structure and culture included in the KB Lottery	KB technology, HR, organisational structure and culture included in the KB cancer research funding handling	KB technology, HR, organisational structure and culture included in the KB cancer information activities and handling	KB technology, HR, organisational structure and culture included in the KB Cancer Patient Support activities and handling
Network - Network and Network partners (e.g. Strategic partnerships, supply chains and others (Physical, Digital, Virtual)	KB network partners involved in the KB lottery	KB network partners involved in the cancer research funding BM	KB network partners involved in the cancer information activities	KB network partners involved in the Cancer Patient Support activities
Relation(s), Relationship(s) (e.g. Physical, Digital, Virtual relations, personal, peers) (Physical, Digital, Virtual)	KB relations inside the KB lottery BM	KB relations inside the KB research funding BM	KB relations inside the KB information	KB relations inside the KB Cancer Patient Support BM
Value Formula (Profit formula - Both turnover structure, Cost structure and revenge floe and other value formula.) (Physical, Digital, Virtual)	Price of lottery - cost of developing and running the KB Lottery	Price of newspaper - cost of developing, running and distributing the KB research cancer funding	BM Price of KB information - cost of developing, running and distributing the KB information BM	BM Price of KB information - cost of developing and running KB Cancer Patient Support BM

Table 5.2 Generic approaches to business model portfolio grouping

Generic approaches to a business model portfolio grouping (each can be physical, digital or virtual)	Core approach related to BM portfolio grouping
Value proposition/s (products, services and processes) that the business model portfolio offers	What are our overall value propositions this group of BMs offers?
Users and customers (users, customers, that the business model portfolio serves)	Who does the BM portfolio serve with this group of BMs – segments, target group?
Value chain functions (internal) that the business model portfolio uses	What overall group or mix of value chain functions do we use to produce this group of BMs?
Competences (technologies, human resources, organizational systems and culture) that transform businesses' inputs into value for customers, users, network partners, machines, employees (outputs)	What are our general competences used for this BM portfolio? These BMs are, for example, produced on the same machine/s, by the same human resources, by the same organizational system, by the same culture...
Network: network and network partners (suppliers and other network partners)	What are our general networks used to operate this group of BMs?
Relation/s (e.g. physical, digital and virtual relations, tangible and intangible)	What are our general relations used for this BM portfolio?
Value formula (profit formulae and other value formulae)	What are our general value formulae used for this group of the BM portfolio?

of KB (Lindgren 2012), a BM portfolio grouping with this point of entry or approach (Table 5.1).

This "triggered" our research to investigate how many BM portfolio grouping forms could be possible. In turn, this research resulted in our finding seven BM portfolio grouping forms or viewpoints to BM portfolios in a business (Table 5.2).

As can be seen, the BM portfolio indicates seven different viewpoints, which we will discuss in later chapters and in Part 2 of this book.

6

The Business Model Ecosystem Approach

Peter Lindgren

Abstract

There is much knowledge about business models (BMs) (Zott and Amit 2009; Zott et al. 2010, 2011; Fielt 2011; Teece 2010; Lindgren and Rasmussen 2013) but very little knowledge and research about business model ecosystems (BMESs) – those "ecosystems" where the BMs really operate and work as value-adding mechanisms, objects or "species". How are these BMESs actually constructed? How do they function? What are their characteristics? And how can we really define a BMES?

There is until now not an accepted language developed for the BMES nor is the term "BMES" generally accepted in the BM literature. This chapter intends to commence the journey of building up such language based on case studies within the windmill, health, agriculture and fair business model ecosystems – the upperpart of the vertical butterfly (Rasmussen, Saghaug and Lindgren 2014; Lindgren 2016b). A preliminary study of "as-is" and "to-be" BMs related to these BMESs present our first findings and preliminary understanding of the BMES. The chapter attempts to define a BMES and its dimensions and components. Every business model is part of or offered to one or more business model ecosystems (BMESs) (Lindgren 2016b). The BMES is where the business BMs operate and "exchange" their value proposition but it is also where the "to-be" BM can be presented in an early stage version – a Beta version or a prototype. The BMES is therefore a different term than a market, an industry, a cluster or a sector, as we will verify in this chapter. In this context we build upon a comprehensive review of academic business and BM literature together with an analogy study to ecological ecosystems and ecosystem frameworks. We commence exploring the origin of the terms "business", "BM" and "ecosystems" and then relate this to a proposed BMES framework (Lindgren 2016b) and the concept of the multi BM framework (Lindgren and Rasmussen 2013).

6.1 The History of the Business Model Ecosystem (BMES)

The first discussion of the business model ecosystem (BMES) can be traced back to an academic article in 1934 (Bloggs 1934, cited in Fielt 2011). However, the concept never really gained wide acceptance until Fielt in the the mid-1990s again raised the question – "How can a BMES be defined?" (Fielt 2011). Fielt commented that:

> The term "Business Ecosystem" was originally used and introduced by Moore (Moore 1993) in his Harvard Business Review article, titled "Predators and Prey: A New Ecology of Competition". Moore defined "business ecosystem" as:
>
> "An economic community supported by a foundation of interacting organizations and individuals – the organisms of the business world. The economic community produces goods and services of value to customers, who are themselves members of the ecosystem. The member organisms also include suppliers, lead producers, competitors, and other stakeholders. Over time, they coevolve their capabilities and roles, and tend to align themselves with the directions set by one or more central companies. Those companies holding leadership roles may change over time, but the function of ecosystem leader is valued by the community because it enables members to move toward shared visions to align their investments, and to find mutually supportive roles."

Moore used several ecological metaphors, suggesting that the business could be regarded as embedded in a (business) environment, that it needs to coevolve with other businesses, and that "the particular niche a business occupies is challenged by newly arriving 'entrants'" (Porter 1985) or potential exit businesses. Moore (1993) further argued for defining the ecosystem as related to the business level and not to the business model level (Skarzynski and Gibson 2008; Osterwalder and Pigneur 2010; Osterwalder 2011; Lindgren and Rasmussen 2013), meaning that business ecosystems should be defined as they related to the highest level of a business and as an ecosystem of businesses or for businesses.

DeLong (2000) defined business ecology as "a more productive set of processes for developing and commercializing new technologies" that is characterized by "rapid prototyping, short product-development cycles, early test marketing, options-based compensation, venture funding, early corporate independence".

Many have tried to define a group of businesses as, for example, a cluster (Porter 1998):

> a geographical location where enough resources and competences amass reach a critical threshold, giving it a key position in a given economic branch of activity, and with a decisive sustainable competitive advantage over other places, or even a world supremacy in that field (e.g. Silicon Valley, Hollywood, Italian clusters) (Dópglio 2011), Danish Wind Valley (Monday Morning 2010; Genoff 2010).

or a sector – Langager (2010) comments on the difference between industry and sector:

> The terms industry and sector are often used interchangeably to describe a group of companies that operate in the same segment of the economy or share a similar business type. Although the terms are commonly used interchangeably, they do, in fact, have slightly different meanings. This difference pertains to their scope; a sector refers to a large segment of the economy, while the term industry describes a much more specific group of companies or businesses.
>
> A sector is one of a few general segments in the economy within which a large group of businesses can be categorized. An economy can be broken down into about a dozen sectors, which can describe nearly all of the business activity in that economy. For example, the basic materials sector is the segment of the economy in which business deal in the business of exploration, processing and selling the basic materials such as gold, silver or aluminum which are used by other sectors of the economy.
>
> Each of the dozen or so sectors will have a varying number of industries.... For example, the financial sector can be broken down into industries such as asset management, life insurance or as e.g., northwest regional banks. The Northwest regional bank industry, which is part of the financial sector, will only contain businesses that operate banks in the Northwestern states – a geographical approach.

An industry, according to (Langager 2010), on the other hand, describes a much more specific grouping of businesses with highly similar business activities. Essentially, industries are created by further breaking down sectors into more defined groupings.

Porter (1985) defined and agreed upon the term industry as referred:

> to the environment and the forces close to a business that affect its ability to offer its value propositions to customers and make a profit.

6.1.1 The "Barriers" or "Borders" of BMES Markets, Industries, Sectors and Clusters

Porter argued that a change in any of five forces – buyers, suppliers, new entrants, substitutes and exit and entry barriers – normally would require that a business had to re-assess "the marketplace" given the overall change in industry formation. The overall industry, according to Porter, does not imply that every business in the industry has the same value formula (Lindgren and Rasmussen 2013) as businesses apply their business models differently.

The industry could in this sense be regarded as equivalent to a BMES – however, it must still be taken into account that Porter's argument concerns business operating in an industry and not businesses operating with one or more business models (Markides and Charitou 2004, Markides 2008, 2013; Casadesus-Masanell and Ricart 2010; Lindgren and Rasmussen 2013). Therefore – according to our findings – Porter may be lacking more or less some fundamental dimensions of a BMES – the value chain functions, the competence, the value formula and not least the relations of the BMES. Further, most cluster, sector and industry frameworks come out of a geographical and physical notation – "thought world" (Dougherty 1992). Porter argued that clusters and industries help productivity, boost innovation and encourage new businesses to evolve. Porter also claimed that businesses' geographical proximity, their close competition with each other and the growth of specialized suppliers and production networks around them made a winning combination.

However many clusters and industries globally seem to be ailing these days – like many ecosystems in biology today – for example, because they are victims of low-cost competition, or in biological ecosystems they are "squeezed" out of their ecosystems by "smarter" species that have adapted to change in the fundamental conditions of the ecosystem with different wants, needs and demands. They "play" a "different model" for survival and growth.

In Como, Italy, for example, an old cluster of silk businesses had for a long time been ailing, as was an old wool cluster around Biella together with the Castellanza cluster. Globalization – a typical changer and influencer of the BMES's basic conditions – had simply made clustering and the formation of industries in this area far less certain – perhaps no longer meaningful.

Business today seems not to be able to protect itself and hide behind borders any longer – the barriers and borders of clusters, sectors or industries as Porter proposed previously (Porter 1985). More open trade, improved transport links and the internet among other explanations mean that bunching together in a cluster, sector or an industry no longer offers strong defence against, for example, cheaper foreign rivals – or business with different BMs. Italy's medium-sized industrial businesses, for example, must adapt to the threat from China and the benefit they previously got from being bunched together in a cluster seems to be weakening (Helg 1999).

Fragmentation of production, value chains and outsourcing abroad are clear signs that businesses have become less competitive, are weakening the networks on which their clusters were built and may even face destroying their previous competitive advantage by clustering or acting as if clustering, sectors and industries still exist.

Successful BMESs in the future may have to be established and look different from those we know of in the past. The approach to the term "BMES" and our view of BMES may have to be seen differently from previous terms like industry, sector and cluster, surrounded by and related to physical and geographical borders. Context borders and approaches might be giving us different and even better strategic advantages than previous terms and "thought worlds".

A deeper and new understanding of the BMES could therefore maybe give us some different and new answers as to why some BMESs are successful and others not – and why a BMES terminology that is more context based could be valuable to future BMI and business model innovation leadership (BMIL) (Lindgren 2012).

6.1.2 The "Barriers" or "Borders" of BMES

Porter introduced the terminology of "barriers" related to industries. In a BMES context we propose to increase this terminology as not just defined as related to physical and geographical barriers surrounding the BMES – but also as related to the digital, virtual and, maybe even more important, the perceptual barriers of the BMES. We propose that barriers in a BMES are context based and really dependent on "who is seeing and sensing" the barriers – or "borders" of the BMES. A BMES formation – we propose – can be much wider than Porter's industry and cluster term – and even cross or mix previous traditionally defined cluster and industry barriers. We claim that this can be an important explanation of why clusters, sectors and industries are suffering today – and some even vanishing – because they try to protect themselves behind barriers that really no longer exist, other business do not see – except in their or others' (governments', societies' or even academics') perceptual picture, viewpoint and mental mindset.

The threat of substitute BMs, the threat of established rivals, and the threat of new entrants – **the three forces of horizontal competition** – and the bargaining power of suppliers and the bargaining power of customers – **the two forces from 'vertical' competition** – have previously (Porter 1985) been regarded as deciding the "BM organization in the industry" – in our term the "BMES culture" – according to Porter, the degree of rivalry between businesses' BMs.

However, as we have seen, previous cluster, sector and industry terminologies were very much defined as related to **the business** and **a single business** – whereas the BMES terminology is related to the BM and the manifold of BMs that a business really has and potentially can create. As we argue, businesses have more than one business model (Lindgren and Rasmussen 2013) and business are seldom represented with all their BMs in one BMES, but with "parts of the business" – one or more BMs – in one BMES and other BMs in different BMESs. Therefore, we can say:

> A business model ecosystem represents more business models from more businesses.
>
> A business is seldom represented in just one business model ecosystem but is more often represented by different BMs in more business model ecosystems.

Figure 6.1 shows a conceptual model of one BMES, with a business offering some of its BMs to the BMES – the unbroken-line triangle – and the dotted lined triangles representing potential BMESs that the business is not yet part of.

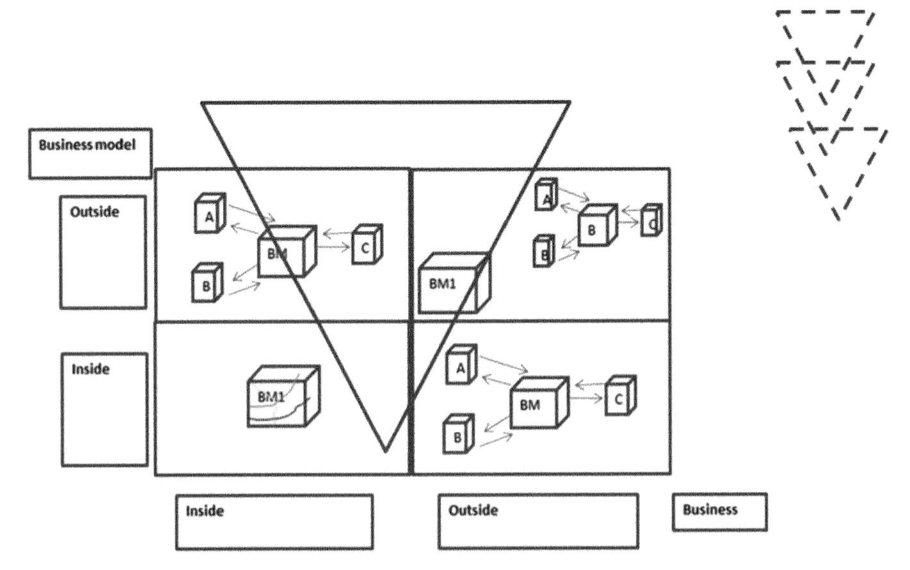

Figure 6.1 Business models and business model ecosystems.

Source: Lindgren and Rasmussen 2012.

6.1.3 Energy in a BMES

The flow of energy through any ecosystem is classically considered as its primary driver according to Lindemann (1940). The flow of energy in an industry, sector and cluster has not yet been fully verified – however some claim that profit is the main driver of any business and, thereby, industry (Max 1867). Lately we have seen that many business ecosystems' real drivers seem to be related to value other than profit (Amidon 2008). In our BMES research we found that the flow of value is one driver of BMES (Amidon 2008, Allee and Schwabe 2011, Lindgren and Rasmussen 2013). However, we found that there may be more drivers to BMES but that profit and also other values seemingly play fundamental roles in any BMES's, business's and BM's "energy" and their "triggers" to make value, create, capture, deliver, receive and consume.

A "system approach" has earlier allowed detailed studies of ecosystems energy and material flow (Odum 1953). A value stream analysis of a BM (Allee and Schwabe 2011) also allows a preliminary study of some of the BMES's value flows (OMG 2015). We claim that values are exchanged through BMESs' internal tangible and intangible relations – and also between BMESs' external tangible and intangible relations. The last we note here as a hypothesis as we have not yet been able in large scale to verify empirically value stream flow between different BMESs. Research (Amidon 2008; Russell 2011), however, claims this is the case.

6.1.4 Business Model Innovation in a BMES

The different BMs participate together in BMES to create, capture, deliver, receive and consume (Lindgren and Rasmussen 2013) value, which also sets the competence and capabilities of any BMES but at the same time also – we claim – the limits of business model innovation (BMI) and potential of BMI in a BMES. This is why some businesses take out their BMs from some BMESs and offer them to other BMESs (Chesbrough 2007) – as they consider some BMESs more sustainable and valuable than other BMESs in the future. For example, some fossil energy businesses in early 2000 slowly began to move from the fossil BMES and enter renewable energy BMES (EON, Shell, Statoil, Dong). IBM also showed this trend by leaving the personal computer BMES and focusing on the service BMES.

The amount of competence inside each BMES's BMs and the amount of value flow from BMs in and out of a BMES – we claim – sets the limits of the BMES's BMI competence, capability, growth and even survival potential. It is vital to any BMES to know about its competences and it is essential to any

BMES to receive value, be able to capture value – preferably new value – and also to be able to consume the value offered. However – and this has not yet been focused upon much in research – any BMES also over time has to be able to relate and deliver value to other BMESs. Very few BMESs over time can stay as a lonely island – an isolated BMES. BMESs need to relate and interact with other BMESs otherwise they will be challenged.

6.1.5 The Business Model Ecosystem Relation Axiom

The flow of value in and out a BMES can be mapped in any BMES and its BMI processes (Lindgren and Rasmussen 2013). Therefore it is important to view any BMES from different "perspectives", which Figure 6.2 illustrates.

Figure 6.2 shows a model of value flow from the different viewpoints of a BMES:

Quadrant 1 – Internal to the individual BMES – A part of a BM's value flow inside a BMES – an example is the different business BM value flow in windmill BMES.

Quadrant 2 – BMESs vertically related – BMESs related as suppliers and customers to each other in an "upstream" and "downstream" value flow – an example is the BMES value chain in Energy BMESs – the coal BMES to the electricity BMES to the household BMES.

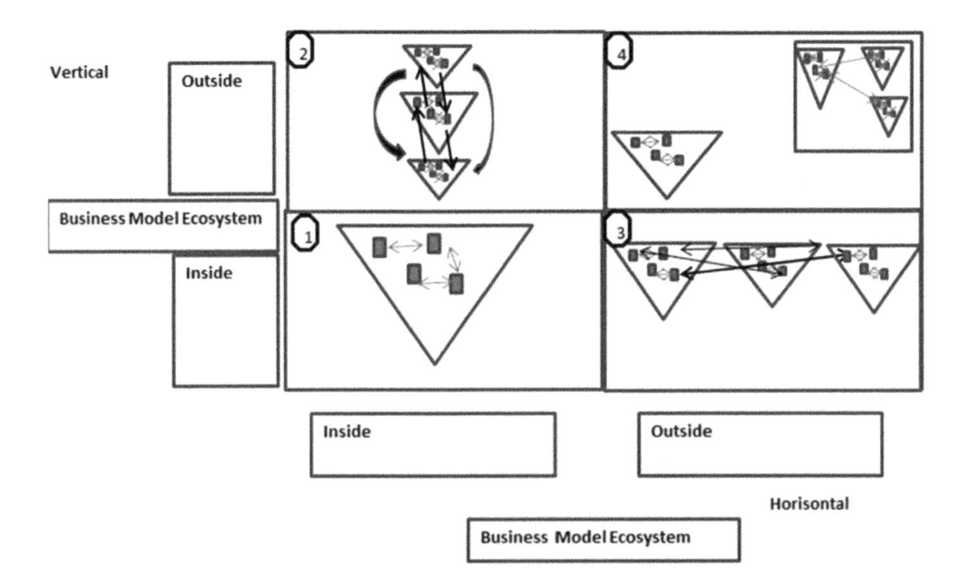

Figure 6.2 BMES relationship axiom inspired by Lindgren and Rasmussen 2013.

Quadrant 3 – BMESs horizontally related – BMESs related as "colleges" in related BMESs – an example is oil, gas and solar electricity in energy production.

Quadrant 4 – BMESs not related – BMESs that are not related to others and make no value exchange. Examples are the windmill BMES and the circus BMES.

Any BMESs are highly dependent, influenced and related to both negative and positive values and value streams from other BMESs. However, value cannot flow between BMESs without one or more relations being created between the different BMESs. This also means that potential value of a BMES cannot be transferred and used in another BMES without relations being established. The study of value flow and relations inside and outside BMESs thereby becomes important to focus on – to verify there are relations and value transfer through the relations – and which BMESs these occur between. A BMES's relations and its BMs' relations to other BMs in different BMESs are fundamental to map carefully to understand the status of a BMES and its potential to BMI. Otherwise it will be nearly impossible to understand the construction and context of a BMES and its growth, survival and potential development.

6.2 Design/Methodology/Approach

The methodology applied in this chapter is structured around deductive reasoning. First, a theoretical background of BMES theory on each dimension of a BMES is presented to provide a foundation. To verify the existence of the dimensions of the BMES and the usability of the BMES, four BMES cases are presented – Danish Energy, Danish Renewable Energy, Suppliers to Danish Energy and HI Fair. To "stress test" the generic use of the BMES framework, the cases represent four very different BMESs with different contexts of BMES dimensions and components. All cases were chosen to exemplify the concept of the BMES in different stages of a BMES life cycle right from construction of a "to-be" BMES to operating as an "as-is" BMES and then a BMES that has lain down to die, prepared to vanish from the scene.

The information and data from the cases were gathered through active participative research (Wadsworth 1998) carried out over seven years in the EU FP 7 IOT project Neffics (Neffics 2012) (2008–2013) and EU Wind in Competence project (2011–2014). Based on these cases supplemented with other empirical cases and tests, a final approach to a definition of the BMES concept is formulated. This is discussed and illustrated in the following paragraphs.

6.3 Characteristics and Dimensions of a Business Model Ecosystem (BMES)

An ecosystem is traditionally regarded as "a community of living organisms" (plants, animals and microbes) in conjunction with the non-living components of their environment (things like air, water and mineral soil), interacting as a system. A BMES is proposed analogically as a "community of living BMs" where different businesses offer their "as-is BM" and develop their "to-be BM" in conjunction with the BMES environment (things like technologies, human resources, organizational structure and culture). In this context and in our approach, BMs that are under construction are also "living" BMs in the BMES as these use energy and competences of the BMES in innovating these "to-be" BMs.

We distinguish here from other frameworks (e.g. Porter) by focusing on the BMs and not the business as forming the BMES. We argue that businesses offers their BMs to the BMES – but very seldom their total number of BMs and thereby their total business. In our research (on the windmill BMES, valve BMES, fair BMES, building BMES, furniture BMES, food BMES food tech BMES and energy BMES) we found that businesses seldom offer all their BMs in just one BMES. Businesses most often spread their BMs to more BMESs – to gain more business, spread risk strategically or because of other reasons. Our research showed that business who offer all or nearly all their BMs to one BMES often face a large strategy risk and are easier to put under value and cost pressure by customers, suppliers and competitors. The strategic best practice saying "stick to your core business" (Abell 1980) is therefore maybe not fully true in all business contexts because the business can be strategically trapped in one BMES by doing so. The saying "focus on your core competence" (Prahalad and Hamel 1990) can be true, when a business offers the same value proposition to more BMESs than one – but can be strategically risky if BMES context bases change.

We distinguish most industry, sector and cluster research and approaches from the BMES approach, as they do not consider and include the "to- be" BM as part of the BMES but what they call a market (Kotler 1983), industry (Porter 1985) or cluster (Porter 1985). We argue that "to-be" BMs are an equally important part of and valuable to any BMES or to many BMESs as there are, for example, customers, suppliers and value propositions that are "flowing" into and out of the BMES and thereby strongly influence the BMES, although these BMs are not fully developed. As an example, we found that "to-be" app and new gaming software development in the Silicon Valley incubation environment are influencing the "as-is" BMs in the app and software BMESs and some of these "to-be" BMs are even "traded" before final launch – even at idea and concept phase.

We acknowledge that many businesses and societies put their primary focus on – and borders around – the BMES's "as-is" BMs – but we point out that this is not giving the full picture and understanding of all BMs, dimensions and characteristics of a BMES. The "to-be" BMs and the proposed "to-be" BMs indeed influence and "value" the rest of the BMES's BMs. Businesses use tremendous resources and energy from the BMES and even other BMESs to carry out their BMI. The BMESs also use energy to protect their "as-is" BMs from "to-be" BMs. "To-be" BMs can be serious and important drivers in the change of "as-is" BMs in the BMES and can also be the source – and give energy – to changing the organizational system and whole culture in a BMES, even in vertically and horizontally related BMESs. Amazon, iTunes and Netflix are just some examples of businesses whose BMs have influenced BMESs that are full of vitality in retail, music and film. "To-be" BMs can disrupt BMESs and sometimes be the drivers to revitalize existing BMESs and related BMESs. "To-be" BMs can naturally be the driver to the establishment of new BMESs, of which Second Life, World of Warcraft and the Tinder Box Festival in Denmark (Tinderbox.dk) are examples.

6.3.1 How Can the "Borders" to BMES be Defined?

Physical borders like land, countries and continents have for many years been regarded as the borders to markets, industries, sectors, clusters and even businesses. Digital and virtual borders in cyberspace such as Google Search, Apple iTunes, Blizzard (World of Warcraft, Zynga) Farmvillage, Viasat TV platform and TDC mobile network are just some examples of BMESs which do not follow these borders, but different ones, often independent of the physical world. Some digital and virtual BMES are free to the user to access (Google Search, Wikipedia) – others are not (Disney World Paris, Legoland Billund). In the latter, you have to be a customer to gain access. Digital and virtual BMESs do most often not stick to the physical borders of yesterday; they push us to change our previous understanding of markets, industry, sectors and clusters.

Kotler (1983) described a market as consisting of values offered to customers to fulfil their wants, needs and demands. Markets consist of customers and suppliers who exchange their values (products and services) for money. Market leaders and market followers compete with each other and prevent new entrants entering the market. Kotler also described markets with special demands for value as "niche markets" and those with indifferent demands as "mass markets". These are small BMESs – ecosystems or communities with special or indifferent value demands. The customers' value demands and the supplier's value offers act as borders for "the ecosystem" and the money is the final determinant of whether a market exists or not.

Porter (1985) described it somehow differently. He defined entry and exit barriers – "borders" – to industry: "exit barriers" prevent businesses slipping out of the industry and "entry barriers" prevent substitutes and new entrants slipping into the industry. These are obstacles that make it difficult to both exit and enter an given industry, hindrances – such as capital investment, government regulations, taxes and patents, or a large, established business taking advantage of economies of scale – that a business faces in trying to exit an enter an industry with its BMs. They can also be the lack of competences a business faces in trying to gain entrance to a profession – such as technology requirements, education or licensing requirements, organizational requirements or cultural practice. Because entry barriers protect incumbent businesses and restrict competition in an industry, they can contribute to distortionary value formulae. The existence of monopolies or industry power often aids barriers to entry – and thereby "the borders" of an industry.

Both Kotler and Porter describe "ecosystems", such as special habits, rules and practice ("culture" (Kotler 1983)), B2C markets, B2B markets (Porter 1985), rivalry, cost leaders, niche and focus strategies. However, the business environment seems in many cases only to be true if these borders really exist. We claim that they might not exist any more or are quickly vanishing.

It seems that they have begun to change or have even vanished since the early 1980s especially with the internet pushing and disrupting borders of markets, industries, sectors and clusters. The internet also provides the opportunity to act in physical, digital and virtual BMESs simultaneously or in an integrated way.

So to answer the question "What are the borders to a BMES?" it might be valuable to rethink the term barriers and borders – and instead think of them as context based. In this case we commence our inspiration and draw an analogy with the science of ecology.

> The biotic and abiotic components of an ecosystem have been regarded as linked together through nutrient cycles and energy flows. **A nutrient cycle** is the movement and exchange of organic and inorganic matter back into the production of living matter. The process is regulated by **food** pathways that decompose matter into mineral nutrients. Nutrient cycles occur within ecosystems. Ecosystems are interconnected systems where matter and energy flows and is exchanged as organisms feed,

> digest, and migrate about. Minerals and nutrients accumulate in varied densities and uneven configurations across the planet. Ecosystems recycle locally, converting mineral nutrients into the production of biomass, and on a larger scale they participate in a global system of inputs and outputs where matter is exchanged and transported through a larger system of biogeochemical cycles. (Chapin et al. 2002)

Ecosystems have been defined by the network of interactions among organisms, and between organisms and their environment: the ecosystems are said to be of any size but usually encompass specific, limited spaces (Chapin et al. 2002; Schulze et al. 2005). However, some scientists even say that the entire planet is an ecosystem (Willis 1997; Schulze et al. 2005; Krebs 2009) – indicating that the borders of ecosystems depend on the context and the viewpoint of the viewer(s).

The tangible and intangible dimensions and components of a BMES are proposed as linked together through relations (Amidon 2008; Allee and Schwabe 2011; Russell 2012). Relations "bind" BMs "context wise" in BMES and they are the "channels" – equal to "pathways" in ecology research – in which values are carried from one BM dimension to another. Relations set the borders for how far the value proposition of a BMES's BMs can reach out and potentially exchange values with other BMs – either inside or outside the BMES. Relations are the vital dimension in a BM and a BMES that can carry value – thereby enabling value exchange and fulfilling a value cycle or a value flow.

When BMs in a BMES are related they can potentially exchange value – but there is no guarantee for value flow and value exchange. Value flow and value exchange are dependent on the value cycle taking place, which means that value will be created, captured, delivered, received and consumed. Obviously much can go wrong or not happen in the value flow process. It depends on many things that are equivalent to the nutrient cycle and "energy flow" in a biological ecosystem, the electricity flow in an electrical system or the heating flow in a heating system. In BMES BMI motivation, trust, ownership, technology, people, organizational systems and culture as examples influence whether value flow and value exchange will and can take place. Relation mapping (Amidon 2008; Russell 2012) can help us to understand better and show which BMs and BMESs carry out which value flow. It can also show how values are exchanged (Allee and Schwabe 2011) between BMs – both tangible and intangible values.

Relations between BMs and BMESs can be both tangible and intangible – and therefore it can be rather complex to study and map BM and BMES value flow, connections of tangible and intangible relations – analogous to nutrient cycles and energy flow study. Mapping of relations in and between BMESs can be even more complex when culture and spiritual dimensions are also taken into consideration (Saghaug and Lindgren 2010).

The motivation and incitements in BMESs and between BMESs to relate have until now not been particularly addressed in research (Lindgren et al. 2014) – but they can be studied through the value flows, value transaction and value network mapping in BMI. Our hypothesis is that there can be more sources than motivation.

To motivate, or trigger, a BMI flow – and a valuable BMI flow – it is necessary and vital to any BMES to exchange value through relations and thereby enable the foundation of all BMI – the learning process (Caffyn and Grantham 2003) – in the BMES. It is important – and vital – to BMESs and BMs that knowledge flow and learning loops happen in BMESs and between BMESs. Any BMESs can benefit from "value adding" knowledge and, conversely, can suffer from its lack.

Learning and motivation to learn is therefore fundamental to any BMI (Lindgren et al. 2014). Motivation to learn is therefore an important trigger or driver to commence a value flow and value exchange.

Energy, water, nitrogen and soil minerals are essential abiotic components of any ecosystem. Analogically, competences (technology, human resources, organizational systems and culture) (Lindgren, Taran et al. 2010) embedded in BMES BMs are essential components of any BMES. Competences can be developed and grow – but can also be diminished, shrink and even vanish in a BMES. Competence can simply disappear or leave the BMES as value flows out – as production leaves a BMES (the Como silk cluster), but also as value flows into the BMES (the Silicon Valley Case).

Value that flows into the BMES can, however, also destroy built up competences inside the BMES and its BMS. We found in our research that both value that flows out and value that flows in can be one of the important reasons as to why some BMESs shrink, collapse and even disappear (windmill, textile and furniture BMESs).

The reasons as to why competence leaves BMs and BMES can be multiple. One could be that competence is forced to leave – Western production in textile, furniture, windmill production and many other industries have left for Asia due to a motivation and perception in the businesses involved of lower production cost, access to new markets and maybe a perception of the possibility of creating a better value formula. Thereby the Western production in these BMESs slowly vanishes as they transfer their competences – technology, HR, organizational system and culture – to, for example, Asia. A "single loop" or a "one way" value flow transfers from one BMES to another BMES.

However, these cases do not obviously increase learning and BMI in the BMES, giving away and sharing value with other BMESs – in this case valuable competences. "Double loop" value flow can conversely – if the receivers of the value are able to capture, receive, consume and create new knowledge and deliver value back to the BMES – enable competence development in the first BMES. A BMES can thereby work as a competence-adding mechanism but also its opposite – either by just giving away value and competences or by developing new value and new competences and sharing these with other BMESs. BMES survival is strongly tied to the capability to continually develop and improve competences – by learning and attracting new value.

Competence of a BMES – the sum of all the BMES BM's competences – therefore makes BMES more or less attractive and thereby vulnerable. Competence is therefore without question a vital dimension (Prahalad 1990) in any BMES – however, paradoxically, it is often still a neglected dimension. Many European and Asian BMESs want, for example, to learn from "the Silicon Valley BMES" competences – learn how to innovate new BMs and business, as, for example, Google, Facebook, Apple and Twitter do, and how to become a sustainable BMES. We believe that continuously learning and knowledge sharing together with motivation to learn from other BMESs are important secrets and essentials to the success of "The Silicon Valley BMES". Silicon Valley has understood the importance of relating to and attracting other BMESs or knowledge zones (Amidon 2008).

6.3.2 "Energy" of Business Model Ecosystems

Living ecosystems – and BMESs – require energy to stay alive. BMESs require available energy to stay alive, grow and even be born. Energy can be stored in the competences of the BMES BMs – or in other BMES BMs – they "only" have to be released (Lindgren and Rasmussen 2013).

BMESs require knowledge of how to release the energy stored in the competences of BMES BMs. The oil industry has the competence (technology, HR, organizational systems and culture) to release oil from "deep under" – but it also has the knowledge inside the BMES to know how to release the oil. The knowledge – how to – is embedded in its BMES competences. If the knowledge – how to – was nonexistent in the BMES, the oil could not be "brought up" or would have to be "brought up" by other BMESs from outside.

The earth receives energy from the geothermal energy contained within it and is sensitive to changes in the amount of energy received. Energy is valuable to the earth – but also to any BMES. A BMES receives value from other BMESs – visible or invisible – and develops the basis of this energy – sometimes in interaction with other BMES's BMs. The BMES, however, also develops energy via the interaction between BMs inside the BMES. We propose that the biological ecosystem and the BMES function in much the same way regarding energy development.

Energy is also stored in the competences of other BMES BMs. Living ecosystems like, for example, the Earth, receive energy from the sun – some would say the sun was an ecosystem outside the Earth's ecosystem; others would increase the Earth's ecosystem to also include the sun. We propose this discussion to be context based related to BMESs as they can receive energy from other BMESs – but a judgement on this is made based on who "sees" and from which viewpoint.

There are, however, different forms of energy. Common energy forms, according to Chapin et al. (2002), include the **kinetic energy** of a moving object, the **radiant energy** carried by light and other electromagnetic radiation, the **potential energy** stored by virtue of the position of an object in a force field such as a gravitational, electric or magnetic field, and the **thermal energy** which comprises the microscopic **kinetic** and **potential energies** of the disordered motions of the particles making up matter. Some specific forms of potential energy include **elastic energy** due to the stretching or deformation of solid objects and **chemical energy** such as is released when a fuel burns. Any object that has mass when stationary, such as a piece of ordinary matter, is said to have rest mass, or an equivalent amount of energy whose form is called **rest energy**, though this isn't immediately apparent in everyday phenomena described by classical physics.

We propose that BMESs also have or develop different forms of energy – however this we have not researched yet and define it terminologically.

Our sun transforms nuclear potential energy to other forms of energy; its total mass does not decrease due to that itself (since it still contains the same total energy even if in different forms), but its mass does decrease when the energy escapes out to its surroundings, largely as radiant energy. Therefore eventually, someday, the sun will stop shining and transforming value and energy into its surroundings. BMESs and BMs also transform potential energy – value and competences – to other forms of energy – value and competences. The total "mass" of a BMES or a BM as a result of the value transformation flow does not reduce either, but as in an ecosystem or in the case of the sun the BMES's and BM's mass does decrease when value or competences escape out to other BMESs or BMs – "single loop" value and competence flow – except when the BMES and its BMs receive value and energy from BMESs outside.

Although any energy in any single form can be transformed into another form, the law of conservation of energy states that the total energy of a system can only change if energy is transferred into or out of the system. This means that it is impossible to create or destroy energy. Any competence in any single form – technological, human, organizational system and culture – can be transformed into another form – inside the BMs, into other internal BMs in the BMES or outside to other BMs in other BMESs. This also means that in BMESs it is impossible to destroy value and competences – but value and competences can vanish to other BMs and BMESs – or as we have seen in several of our cases in our researches (Newgibm case research 2006; Blue Ocean case research 2008; WIB 2012, ICI case research 2013; Neffics 2012; SET cases 2014; EV Metalværk 2014), it can rest as hidden values and competences (Lindgren and Saghaug 2012) inside a BM or a BMES.

6.4 Introduction to the Business Model Ecosystem (BMES)

The focus is not on the BM but on the BMES and the dimensions and construction of BMES which any BMs are a part of. Although this is not sufficient to cover the whole BMES theory framework approach as it is just one focus of probably many viewpoints of BMES; it is an attempt to describe a fragmented part of the whole business model environment, research and discussion.

We try to find the dimensions and components of BMES that everybody seems to acknowledge and add those we believe are missing. We try to merge those dimensions which are overlapping and we try to take out those dimensions that are not vital for BMES. From this point of entry, we test our BMES dimensions in four BMES case studies to verify empirically our hypothesis of the existence of seven dimensions of any BMES.

6.5 Dimensions, Concepts and Language of a Business Model Ecosystem (BMES)

From acknowledged academic works and our research work with the dimensions of a business model and business, we found some generic dimensions that support the idea that any BMES could also with preference be defined by seven generic dimensions.

6.5.1 Value Proposition Dimension of a BMES

All businesses we investigated offer values to either BMs inside the BMES and/or to BMs outside the BMES. The BMES value proposition seems to be a "mirror" of the BM's value propositions individually and together inside the BMES. We define these as the BMES value proposition offered to other BMs either as one BM to another or more BMs together as a shared value proposition of the BMES. Value propositions from a BMES can be offered in the form of products, services and/or processes of services and products.

6.5.2 Customers and/or User Dimension of a BMES

A BMES serves customers and/or users (Appendix 1).

> A successful *BMES* is one that has found a way to create, capture, deliver, receive and consume value for its users and customers – that has found *"a way"* to help customers *and users of a BMES* to get an important job done – *"solve pains" and "create gains" for its "users" and "customers"*. "It's not possible to invent or reinvent a BMES without first identifying a clear customer and/or user base".

Here, we draw a distinction between customers and users of a BMES. Customers of the BMES pay with money – "there is no BMES marked – Business of a BMES – if the customers of a BMES do not pay" (adapted from Kotler 1984), whereas users of a BMES pay with other values (von Hippel 2005) than money. Business model theory (Appendix 1) has mainly considered the business model related to customers. However, as we have verified in our research (Lindgren and Rasmussen 2013) users can be highly valuable to a BMES by "paying" with other values (Facebook, Google). Industry, sector and clusters mostly focus on money but do also consider other values as payment to a BMES.

6.5.3 Value Chain Functions (Internal Part) Dimension

Any operating BMES has functions that it has to carry out and which enables the BMES to "offer" the value propositions to its customers and users. A value chain function list including primary and secondary functions of a BMES can be created. Primary functions can be inbound logistics, operation, outbound logistics, marketing and sales, service; and secondary functions – support functions – can be procurement, human resource management, administration and finance infrastructure, business model ecosystem innovation. These do not have to all be present and carried out to have the BMES operating.

Any operating BMES needs to have someone to carry out these functions to enable it to create, capture, deliver, receive and consume a value proposition to and from its users, customers and network. These can either be carried out by its own users, customers, competence and network or by other BMESs.

6.5.4 Competences Dimension

In BMs we have earlier (Lindgren and Rasmussen 2013), inspired by Prahalad and Hamel (1990) divided competences in to four groups – technology, human resources, organizational system and culture. In a BMES we also consider the competence dimension to be technology, human resource, organizational system and culture with the different BMs “pooling” their competences. The pool of these competences forms the “shared competences” available in the BMES.

6.5.5 Network Dimension

We acknowledge that some BMESs sometimes regard themselves as isolated from other BMESs or do not relate to others. We argue that any BMES, whether they want it or not, are in a network of BMESs – and that these networks of BMESs can be physical, digital and/or virtual (Goldman et al. 1995; Whinston et al. 1997; Child and Faulkner 1998; Child et al. 2005; Vervest et al. 2005; Lindgren 2011). We found that the most “successful” BMES is the one that has found a way to create value for its network of BMESs, to help the network of BMESs and/or to get an important job done for the network of BMESs.

Some BMESs mention or communicate openly the BMES network in which they exist and collaborate – others do not. Many BMESs do not understand and often do not acknowledge value which they receive from other BMESs before it is too late and they are in risk of vanishing, or being punished or restricted.

6.5.6 Relation Dimension

Business models are related through tangible and intangible relations (Provan 1983; Provan et al. 2007; Provan and Kenis 2008; Allee and Schwabe 2011) to other business models (Håkansson and Snehota 1990; Amidon 2008; Russell 2012; Lindgren and Rasmussen 2013). Businesses are related through strong and weak ties (Granovetter 1973). As BMESs are a construction of BMs it seems also obvious that these are to be related through tangible and intangible relations – and also with strong and weak ties. BMESs send value propositions to other BMESs through relations and receive value propositions from other BMESs through relations. Relations can be one to one or one to many. Relations can be visible and invisible to humans or machines (Lindgren 2012). Tangible and intangible relations are used in the BMES to deliver and receive values (Allee and Schwabe 2011). BMESs relate their BMs’ value proposition, users/customers, value chain functions, competences and network through relations. Relations are used for creating, capturing, delivering, receiving and consuming values.

6.5.7 Value Formula Dimension

Any BMES uses some kind of formula to calculate the value it offers to its own BMES or other BMESs. The value formula shows how the value proposition delivered is calculated by the BMES. The result of this calculation is a value formula either expressed in money and/or other values.

It has been documented that the BMES operates and is influenced by its BMES environment – external environmental factors. In this chapter, we leave out these external environmental factors (political, economic, social, technical, environmental, legal (PESTEL)), conditions and competitive contexts and environment dimensions, acknowledging that the BMES's external environment is important and critical to its survival and growth. However, we believe that these environmental factors are outputs from other BMESs.

The seven dimensions mentioned in this section of the chapter are equivalent to the overall model we propose to show how any business and business model is constructed (Lindgren and Rasmussen 2013). The seven dimensions we propose should also be considered by any BMES. However, there is a difference between the way businesses want to run their operations in a BMES – seven visionary dimensions of a business – and how a business really runs its operations in a BMES. By mapping empirical data from our BMES case studies to the seven dimensions, we found that business run their BMs differently in a BMES and most businesses have more than one BM in a BMES. In other words, the businesses they described via the seven dimensions are different to how they actually run their business models in the BMES. Some of these business models were close to their original description of the seven dimensions but others were different. This often challenges the survival and growth of a BMES – but it also drives the development, organizational system, culture and vitality of a BMES. If more businesses begin to run their BMs out of "sync" with the BMES's overall vision, mission and the goals of the seven dimensions then the BMES can be challenged and eventually be disrupted, torn apart and vanish.

This places our attention on the "download", "see" and "sense" approach to the BMES using the perspective that BMESs have more BMs that are different as seen in Figure 6.3. We address the importance of continuously investigating BMESs and their BMs and innovation of BMESs to "picture" the distinction between the "visionary model" of the BMES and the BMs of business that are actually carried out ("as-is" BM) and are intended to be carried out ("to-be" BM) in the BMES. Herein, we believe, lays the "seed" to BMESs' survival.

This observation, together with inspiration from Abell's and Hamel's original definitions and framework of "the core business" (Abell 1980) and "the

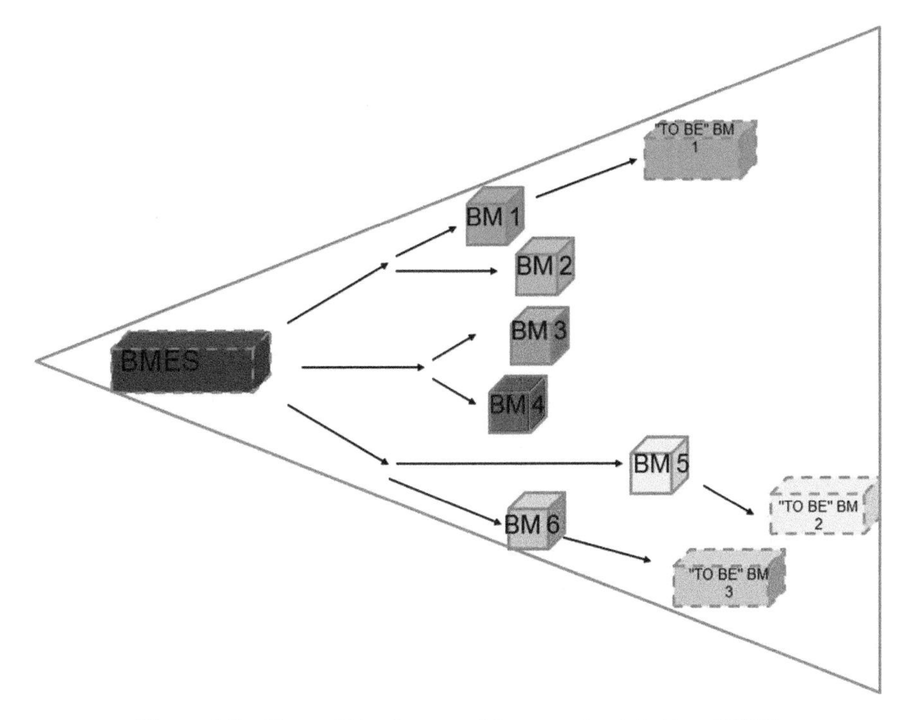

Figure 6.3 The multi business model approach related to a BMES.

core competence" (Hamel and Prahalad 1994), made us draw an analogy with the definition of "the BMES" as the BMES context – and the visionary level states how BMESs are related to the seven dimensions mentioned in this section of the chapter.

The core of the BMES refers, therefore, in this perspective to:

> How a BMES is constructed and intends to operate its "main" and "essential" business related to the seven BMES dimensions – value proposition, user and/or customer groups, value chain (internal functions), competence, network, relations and value formula.

In this context we acknowledge that some BMESs operate without a strong vision, strategy or intention – others not – or that these evolve as the BMES grows, lives and dies.

In our research, we found that many BMESs do not stick strictly to their core business and how they were meant or intended to run and be. They have, in fact, a variety and mix of BMs which sometimes have different value propositions, users and customers, value chains with different functions, competences, networks, relations and value formulae – they cross "the borders" of "the core BMESs". One set of dimensions of a BMES does not always fit all

Table 6.1 Generic dimensions of a BMES

Core dimensions in a BMES (each can be physical, digital or virtual)	Core questions related to dimensions in a BMES
Value proposition/s (products, services and processes) that the BMES offers	What value propositions does the BMES provide?
Customer/s and users that the BMES serves – geographies as well as physical, digital, virtual	Who does the BMES serve?
Value chain functions (internal)	What value chain functions does the BMES provide?
Competences (technologies, HR, organizational system, culture)	What are the BMES's competences?
Network: network and network partners (strategic partners, suppliers and others)	What are the BMES's networks?
Relations(s)	What are the BMES's relations?
Value formula (profit formulae and other value formulae)	What are the BMES's value formulae?

BMs and businesses. This mix of dimensions – which we classify as different BMs – exists and coexists within the core business of the BMES – what we call BMs inside the business – but also exists and coexists outside the BMES. Individual BMs are not necessarily aligned strictly nor have to be aligned to the core business model of the BMES and the seven dimensions of the BMES.

We argue therefore that a BMES's different BMs cannot be explained by just one BM – "the core business model" of the BMES – but would preferably be better explained by different BMs in the BMES – still each with seven dimensions, but with different characteristics. In our research, we found many examples of different BMs operating in a BMES.

As a consequence, we propose that any BMES can be said to have more BMs offered by different businesses – the multi business model approach (Lindgren 2011) – which are more, less or not aligned with "the core business model" of a BMES. However, any of these BMs can be defined as related to an overall generic BMES BM consisting of seven generic dimensions. Each of the seven dimensions of a BMES addresses some core questions in relation to each individual BMES's dimension's characteristics and logic (see Table 6.1).

6.6 The BMES BM's Dimensions and Component Level

Each BMES can be divided into different dimensions and components. We now exemplify the BMES dimensions and components by explaining firstly how each dimension and component in any BMES can be different and then how they can be characterized on a BMES dimension and component level.

6.6.1 The Value Proposition Dimension – "What Value Propositions Does the BMES Provide?" (VP)

BMs are key in understanding the value "offered" in a BMES. However, BMs vary in the BMES related to their different BMs' dimensions – value proposition, users, customers ... the BMES's value proposition is often very complex to understand in detail because it is not static but dynamic over time. The BMES's value proposition is also complex to understand because it is often a mix of shared value propositions offered by more BMs. Therefore, the BMES's value proposition has to be understood from different perspectives, for example of the BMES customer and/or user it is servicing, its network partners, by the context the BMES delivers its value proposition in, the time in which the BMES delivers its value proposition and the "place" where the value proposition is offered by the BMES (physical, digital or virtual). The BMES can be said to be closely connected to the concept of "the BMES's total value and cost to its users, customer and network partners". In this case, staying at the point of entry to a BMES or a BMES's value proposition process over time is strongly related to the user's, customer's and network partner's total perceived value and total perceived cost of the value proposition offered by the BMES. This is why it is incredibly difficult from the outside to measure, read the values and cost of a BMES and how the users, customer and network partners value it, and decide the degree of attractiveness of a BMES.

Classifying the value proposition of BMES is often different for each **user, customer, network over context, time** and **place**.

Inspired by Payne and Holt (1999) we outline four types of values related to values proposed by a BMES.

1. **Use value** – the properties and qualities which accomplish a use, work or service for the users, customers and network.

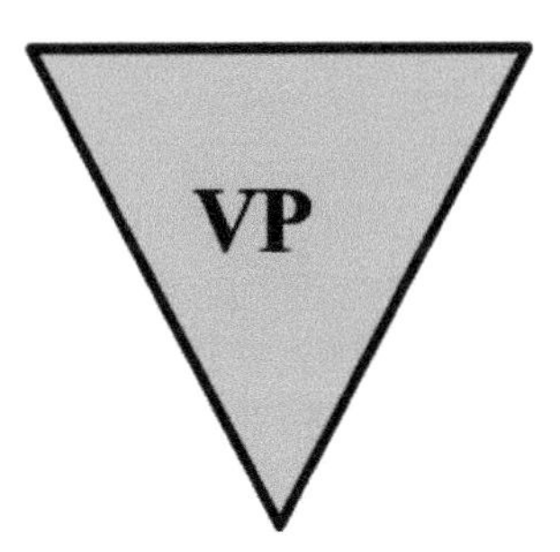

Figure 6.4 The value proposition dimension of a BMES.

2. **Esteem value** – the properties, features or attractiveness which inspire a desire to own the product, service or process in the users, customers and network.
3. **Cost value** – the sum of labour, materials, and various other costs required to produce value for the users, customers and network.
4. **Exchange value** – the properties or qualities which enable exchanging the value proposition for something else that the users, customers and network want.

We found that the list of types of BMES values that solve "the pains and gains" (Osterwalder 2014) of BMES users, customers and network has to be complemented by an overall dimension of the BMES work time vs. life time (Kirkeby 2000, 2003). Time as the factor that defines the BMES's users', customers' and network's personal or BM values of being part of the BMES – the, for example, trade or process related to an overall lifetime value perspective of the BMES – and describes the sum of actions taken in order to find work life-fulfilling and transcend the BM, a value often seen as the driver of the BMES (Tillich 1951; Austin and Devlin 2003; Sandberg 2007).

The value proposition of a BMES has to be measured **before, during** and **after** the BMES exists. This means that a BMES's users, customers and network could trade or collaborate on the different value and cost the BMES offers but also on the value of the relationship that exist in the BMES and between BMESs. The creation, capturing, delivering, receiving and consumption of values from the BMES through its relations are the value creation, capturing, delivery, receiving and consumption of an "inter-BM organizational collaboration business" – a network-based BM business. This is one important value and also an attraction factor, which could be, in this case, a BMI of a "to-be" BMES – when existing BMES's BMs are not enough. The value formula of this can be money to the BMs participating in the BMES (Apple's App Store, YouTube, Food Tech 2014 Fair, Roskilde Rock Festival), but it could also be other values, e.g. learning, supporting a vision, a case (Greenpeace, the Red Cross, a political party). This is in line with research claiming that the value of relationship, activity links, resource ties, and actor's bonds (Håkansson 1982; Axelsson and Easton 1992; Håkansson and Snehota 1995; Ford 2001; Ford et al. 2002, 2003) can be even more important than the value of money for products or services of a BMES. The value of the relationship of a BMES is both an input and an output of the BMES and BMES innovation process, which supports the argument that value and cost of a BMES are not static but dynamic.

As values are created, captured, delivered, received and consumed in a value process in the BMES; BMESs are continuously undergoing change throughout the BMI process or the lifetime of a BMES. The values and cost of BMES relations can be related directly (e.g. profit, volume, safeguard functions) but also indirectly (e.g. innovation, market, scout, access functions). The value and cost functions can further be of a low- and/or high-performing character which is often up to the user's, customer's and network partner's judgement to influence the degree of this value and cost.

The value and cost of a BMES should also be understood as perceived value – benefits and cost (Woodruff 1997; Walter et al. 2001; Lindgren and Dreisler 2002), which means that the real value of BMES can in some cases be neglected in favour of a higher or lower perceived value of the BMES value proposition.

Furthermore, perceived value should not just be related only to each individual BM in the BMES but also to groups of BMs in the BMES – what we propose be called the portfolio level of a BMES. Therefore, it is the user's, customer's, competence's and network's interpretation of "value" and "cost" that is important and not just what "the business of the BMES", its stakeholders (investors, the industry, sector, cluster), society and others think ought to be or are the values and cost of a BMES.

It is therefore very complex when analysing and understanding a BMES's product, service and/or process of value proposition, to analyse all BMs' and stakeholders' values, costs, perceived values and the costs of a BMES. Furthermore, it is important to analyse these over time, during trades or inter-BMES collaborative processes, as values and cost are dynamic and will therefore by definition always change throughout the entire value and cost innovation process and thereby over time. Today no industry, sector and cluster framework has managed to cover and capture value and cost change over time – from different viewpoints. The holistic picture of a BMES value proposition is still very blurred and very complex "to see" but opens up to a whole new way of viewing value contrary to the market, industry and cluster approach.

In summary, any BMES may offer a value proposition – tangible and/or intangible. Value proposition from a BMES can be expressed in value propositions but also in the values of relations. In fact, the values of a BMES can be seen at least from seven different viewpoints, which we comment on in Part 2 of this book.

6.6.2 Customers and Users Dimension – "Who Does the BMES Serve?" (CU)

All BMESs that we researched had users and customers. However, we found that many BMESs do not have customers that pay for the BMES's value

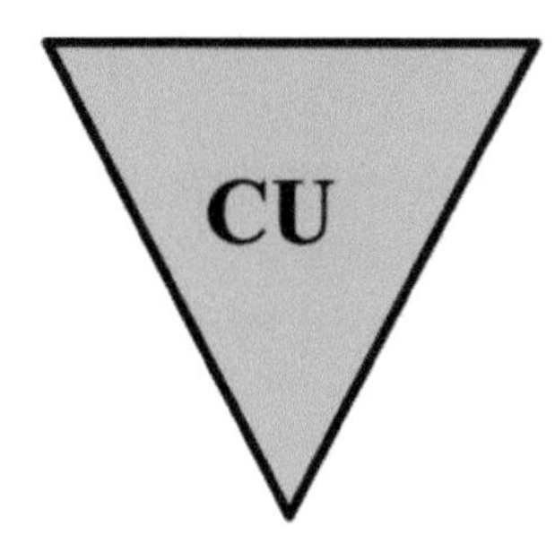

Figure 6.5 The customer and user dimension of a BMES.

proposition. Several BMESs are "just" constructed around users – maybe for a very limited time and a limited topic (Brent Spar Shell 2014), which provides the foundation for the BMES or even for other BMESs with customers related to the BMES – sponsorship, membership, likes, referrals. Facebook, Skype, LinkedIn, Twitter and Google could be examples of such BMES. This indicates that a complete mapping of the BMES BMs can be extremely difficult to establish – also because our research shows that BMs in different BMESs can be users and customers of the BMES in focus at the same time – but in very different contexts.

Our research showed that BMESs built upon users, when growing big in numbers of users, can attract and activate customers from other BMESs willing to buy or pay for value propositions in BMs in the BMES (Facebook, Skype, LinkedIn, Twitter and Google as examples again). Either users start to pay for better performance, advanced use, deeper content, for example, or other customers from other BMESs buy, for example, promotion, data, analytics because there are so many and valuable users in the BM. In these cases, the customers pay for other or different value propositions – or a different BM – as access to, for example, knowledge and learning about the users in the BMES is attractive. Stock buyers placed in a different BMES to Facebook and Alibaba.com BMESs could be an example of this.

6.6.3 Value Chain Functions (Internal) Dimension – "What Value Chain Functions Does the BMES Provide?" – (VC)

All BMESs carry out certain functions to produce the value proposition to the users and/or customers and network partners. Porter's value chain framework was related to an operating business. However, when BMESs start to create a "to-be" BMES there are really no active activities, just wishes and expectations of value chain functions the BMES should carry out. Further, when

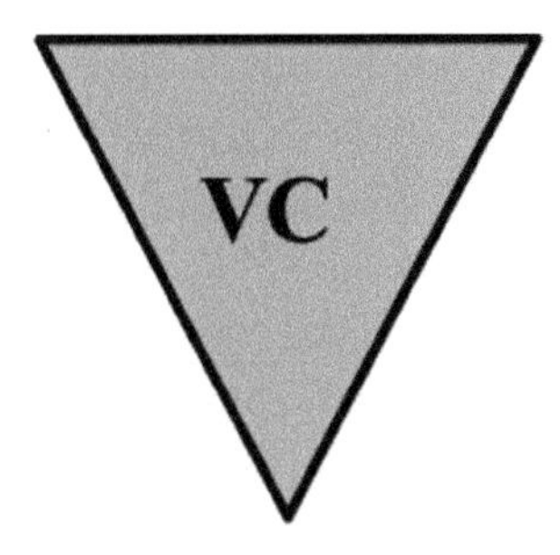

Figure 6.6 The value chain function dimension in BMES.

we observe an operating BMES at a certain moment – in this case, we freeze the picture of a specific BMES – we do not see "running" functions but just functions that are carried out. Value chain functions in our BMES framework represent the value chain functions that have to be carried out or are being carried out within the BMES – internal value chain functions in the BMES. We acknowledge that there are value chain functions outside the BMES but in our framework we only focus on the internal value chain functions of the BM.

6.6.4 Competence Dimension – "What are the BMES's Competences?" (C)

All BMESs rely on and use competences, either from the focal BMES, from BMES network partners or even from BMES customers and users to carry out the value chain functions to be able to create, capture, deliver, receive and consume the value propositions of the BMES. As we have discussed, according to Prahalad and Hamel (1990) competences can be divided to four main categories: technologies, HR, organizational system and culture.

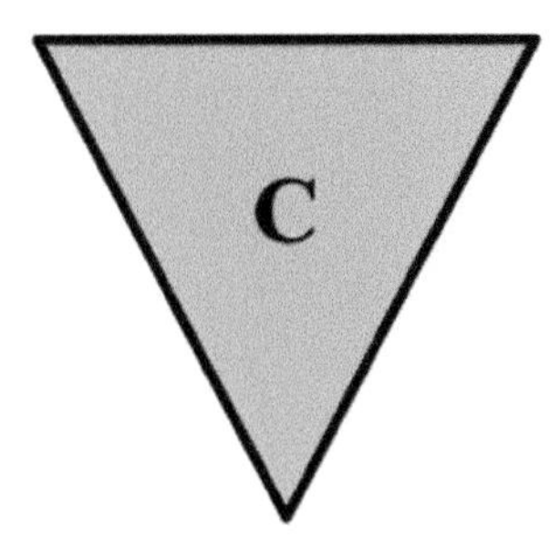

Figure 6.7 The competence dimension of a BMES.

Technologies within a BMES we divided into:

1. Product and service technologies of a BMES
2. Production technology – both "product- and service-production technologies" of a BMES
3. Process technology – that runs and steers the production technologies so that the product, service and production technologies can create, capture, delivere, receive and consume the value propositions of the BMES.

Each BMES has a specific mix, integration and use of product and service technologies, production technologies and process technologies. Sometimes the mix, integration and use of technologies is so unique to the BMES that the competence can be a core competence of a BMES in relation to other BMESs.

Human Resources are "the people" of the BMES placed in the BMs in the BMES.

Organizational system is what the BMES uses to organize the use of BMES technologies, human resources and culture to carry out the value chain functions.

Culture is the "soft" part of the competence dimension. We claim that any BMES has a specific culture.

6.6.5 Network – "What are the BMES's Networks" (N)?

No BMES is a lonely island – at least not for very long. Why? Because if a BMES does not receive value from outside, our research shows that it will slowly shrink and vanish. If it does not offer a value proposition of any kind to another BMES it will not be able to receive value from a long-term perspective. The BMES network thereby becomes vital to any BMES.

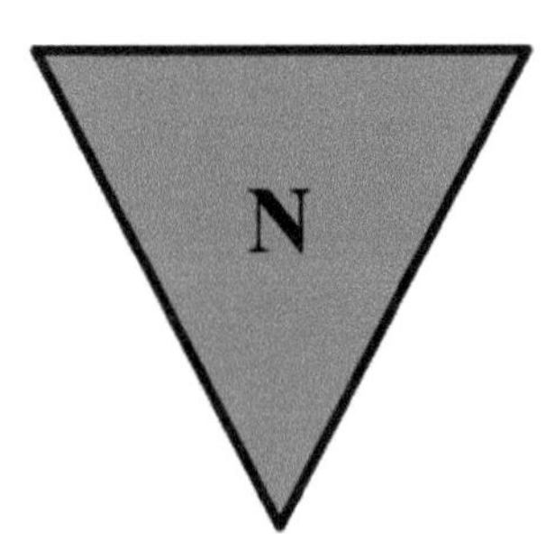

Figure 6.8 The network dimension of a BMES.

6.6.6 Relations Dimension – "What are the BMES's Relations?" (R)

Any BMES relies on relations between BMs inside the BMES. In our research, we, however, found four sets of relations that are of importance to BMESs and should be attended to.

1. The **"inside BMES inside BMs"** area relations – business model relations transferring values inside the BMES BMs.
2. The **"inside BMES outside BMs"** area refers to relations between different BMs inside the BMES.
3. The **"inside BMES outside BMES"** refers to relations between the BMES's BMs outside of the BMES.
4. The **"outside BMES outside BMES"** refers to relations and relation areas where the BMESs do not share a relation to the BMES that are different.

Value propositions and competences of a BMES can be seen from many perspectives as shown in Figure 6.2 at the beginning of the chapter. Value propositions from a BMES can not only be related to products, services and processes of the BMES but also strongly connected to its relations and thereby a result of the relation between BMESs, activity links, resource ties and actor's bonds (Håkansson 1982; Axelsson and Easton 1992; Håkansson and Snehota 1995; Day 2000; Ford et al. 2003). These are all tools which can be used to describe and map relations to and in the BMES.

The creation, capturing, delivering, receiving and consumption of value in a BMES is enabled through these relations (Lindgren 2012). Relations connect the different BMESs' BM dimensions' components and enable the creation, capturing, delivering, receiving and consumption process of value. However, if a BMES is not able or willing to relate and later send and receive value through relations, then the relation has no value, no task – and gives no obvious meaning and value to a BMES.

Figure 6.9 The relation dimension of a BMES.

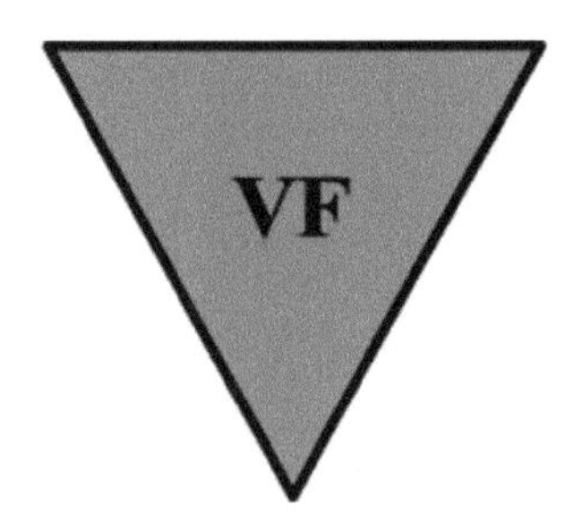

Figure 6.10 The value formula dimension of a BMES.

6.6.7 Value Formula Dimension Component Level – "What are the BMES's Value Formulae?" (VF)

Any BMES will have one or more value formulae, which can be expressed in either a monetary and/or in a non-monetary value formula. We found that the term "profit formula" is too narrow a terminology to express the formula by which BMES calculates the value formulae of a BMES. Our research showed that many BMESs and their BMs are not focused, or, better, are not exclusively focused on profit but instead on other value formulae of the BMES. They "calculate" on other value formulae and to get a full understanding of why BMESs exist and are innovated it is definitely necessary to include other value formulae. We propose profit formula as one of many value formulae that can be the "calculated" output of a BMES.

Having proposed that the seven dimensions of the BMES exist, it enables us to complete the concept and picture of the generic BMES, which we believe can be expressed with the same generic model and questions as proposed in the B-star model (Figure 6.11).

However, we discovered that the seven dimensions form a BMES cube with the "IN IN" relations inside the BMES, as shown in a sketch model in Figure 6.12.

The 2D version is very helpful when working on a BMES dimension level and a 3D version would be helpful when working on a BMES in a BMES relation axiom level. Both presentations would be helpful when working on BMI of BMES.

6.7 BMES Cases

6.7.1 Case 1 – Danish Energy BMES

The Danish energy market can be considered in a certain context as a BMES. Oil (Mærsk, Statoil, ELF, Shell, Dong, Q8, OK, etc.), coal (Dong, Neas, etc.),

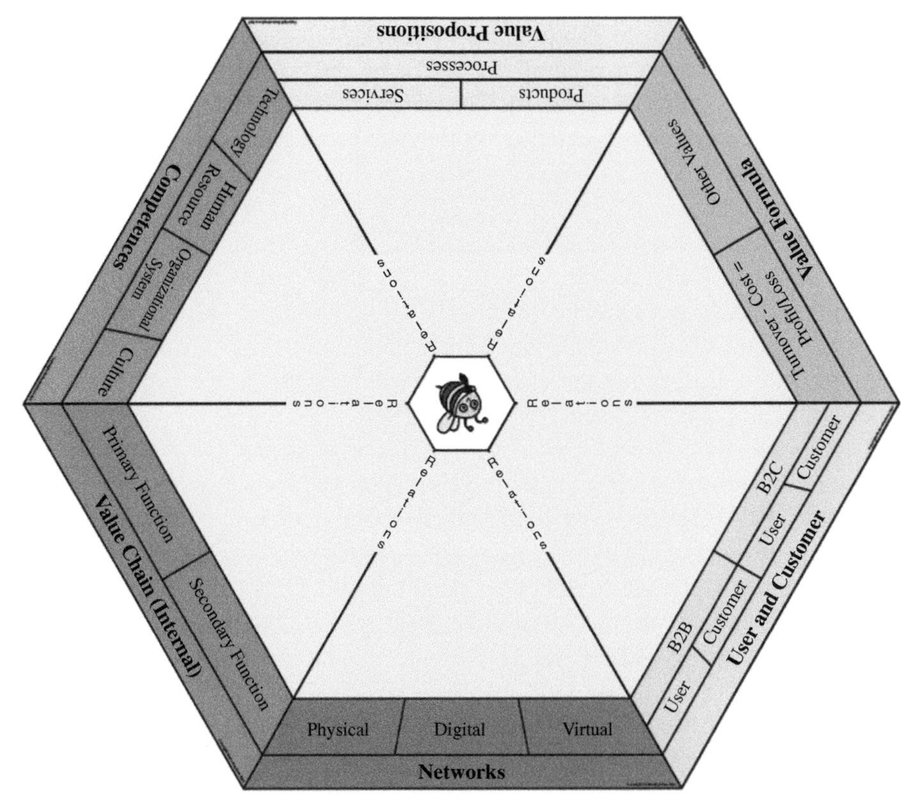

Figure 6.11 The seven proposed dimensions of the BMES. Beestar: Source and ©The BeeBusiness.

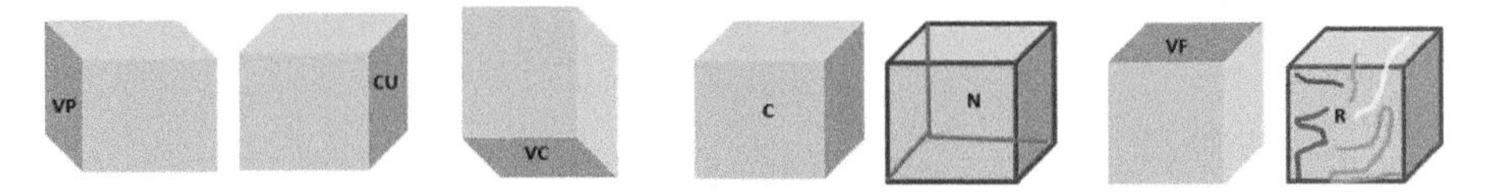

Figure 6.12 The seven dimensions of a BMES presentation.

gas (Dong, Praxair, Kosan, etc.), Biogas (EON, Blue Planet, Maabjerg, etc.), solar (Dansk Sol Energy, etc.) and electricity from windmills (Dong Energy, Watenfall, Neas) are considered as major energy forms in the Danish.

Energy BMES. As can be seen different businesses operate in the BMES and some businesses even operate with more than one BM in the BMES (Dong, Shell, EON, etc.).

Denmark has considerable sources of fossil energy – oil and gas from the North Sea. The production of oil fell from 523 PJ in 2010 to 470 PJ in 2011.

Consumption of oil fell from 315 to 306 PJ (Dansk Statistik 2012). Denmark expects to be self-sufficient in oil until 2050. The production of natural gas fell from 307 PJ in 2010 to 265 PJ in 2014. Consumption fell from 187 to 157 PJ. However, gas resources are expected to decline and production may fall below consumption in 2020, making imports necessary. Politically there is a major wish to exchange natural gas ("black gas") with Biogas but Biogas only took 3 per cent of total gas consumption in 2014 (DWI 2014). The Danish government have announced that the aim is to have "black gas" exchanged for more "green gas" so that Denmark can save more CO_2 and become more independent of fossil gas (Danish Ministry of Climate and Energy 2011). Businesses that operate in the Biogas market today are several private biogas producers together with companies including EON and HMN.

A large proportion of electricity is still produced from coal but a growing part by wind turbines, which met about 39 per cent of electricity demand in Denmark by 2014 (https://en.wikipedia.org/wiki/Wind_power_in_Denmark). To encourage investment in wind power, families (customers) were offered a tax exemption for generating their own electricity within their own or an adjoining commune. While this could involve purchasing a turbine outright, more often families purchased shares in wind turbine cooperatives which in turn invested in community wind turbines. By 2004 over 150,000 Danes were either members of cooperatives or owned turbines, and about 5,500 turbines had been installed, although with greater private sector involvement the proportion owned by cooperatives had fallen to 75 per cent.

In February 2011 the "Energy Strategy 2050" was announced by the Danish government with the aim to have Denmark become fully independent of fossil fuels by 2050 (Danish Ministry of Climate and Energy 2011). The Danish government target is to have 50 per cent wind power in the electricity system by 2020 – a major change in the relative balance between energy sources in the Danish BMES.

Denmark's electrical grid is, however, connected by transmission lines to other European countries (other BMESs) – Norway, Sweden, UK and Germany and has thereby, according to the World Economic Forum, the best energy security in the EU – but is also heavily influenced by these BMESs. In Table 6.2 a description and analysis of the Danish Energy BMES are presented.

Coal power provided 48.0 per cent of the electricity and 22.0 per cent of the heat in district heating in Denmark in 2008; and in total provided 21.6 per cent of total energy consumption (187PJ out of 864PJ) and is based mainly on coal imported from outside Europe (other BMESs). Businesses operating in this market are primarily Dong Energy, Watenfall and others).

Table 6.2 Fossil fuel consumption in Denmark

Energy in Denmark						
Year	Capita (Million)	Prim. Energy (TWh)	Production (TWh)	Export (TWh)	Electricity (TWh)	CO_2 -Emission (Mt)
2004	5.40	233	361	117	35.8	50.9
2007	5.46	229	314	64	36.4	50.5
2008	5.49	221	309	54	35.5	48.4
2009	5.52	216	278	43	34.5	46.8
2010	5.55	224	271	42	35.1	47.0
2012	5.57	209	244	19	34.1	41.7
2013						
2014						
change 2004 to 2014	+3.7%	–10%	–32%	–84%	–4.7%	–18%

Denmark has also two geothermal district heating plants, one in Thisted, founded in 1988, and one in Copenhagen, operating from 2005. They produce no electricity.

In 2012 Denmark reached its year 2020 governmental goal of installing 200 MW of photovoltaic capacity. As of 2013, the total PV capacity from 90,000 private installations amounts to 500 MW. Danish energy sector players estimate that this development will result in 1,000 MW by 2020 and 3,400 MW by 2030.

In the model of the Danish Energy BMES (DEB) it is possible to see registered operating business models.

6.7.2 Case 2 – Danish Renewable Energy BMES

The Danish energy BMES as sketch in a model in Figure 6.13, could also be seen in another context where the focus is just on the renewable energy BMES, as seen in Figure 6.14. The renewable energy BMES in Denmark consists of electricity from windmills (Dong, Watenfall, Neas), solar energy (Dansk Solenergy, Energy Midt, private households, etc.), Biogas (EON, Sydenergi, etc.), geothermal energy (Thisted Termical Energy, etc.), and blue energy based on algae (Blue Energy, Folum, etc.) as seen in Figure 6.13.

The market volume of the BMES for renewable energy in Denmark is of course smaller than the total energy BMES in Denmark. Further, some of the minor business models in the energy BMES suddenly become bigger and even large players if we change the context to now only considering the renewable energy BMES.

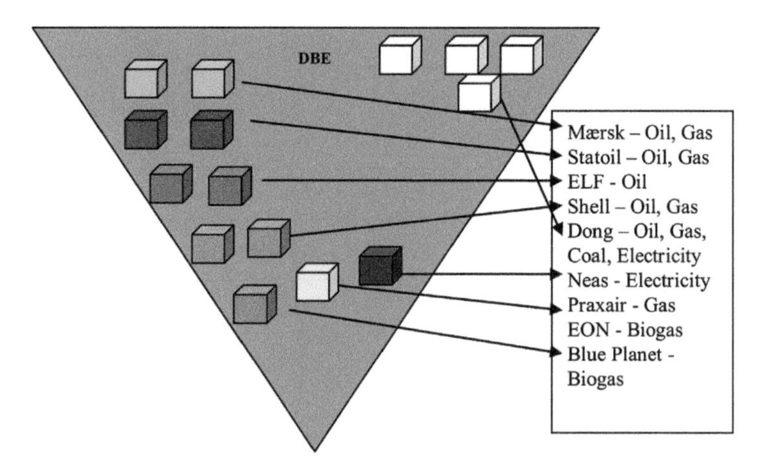

Figure 6.13 Danish energy BMES with elected BMs of operating businesses.

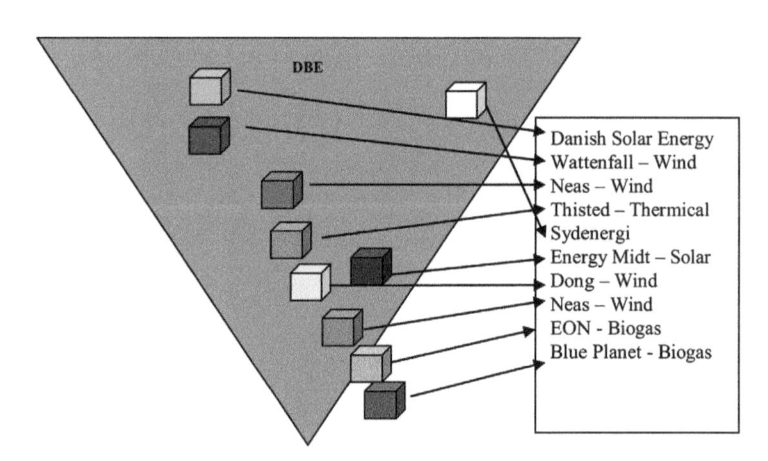

Figure 6.14 Renewable energy BMES in Denmark.

Also interesting is that the numbers of "to-be" BMs and the degree of innovation increase in the renewable energy BMES compared to the energy BMES. Some universities and GTS institutions are now actors in the BMES with a different value formula than money – namely research and learning as a focus. Also several municipalities, regions (Denmark is divided into five regions) and even the state government are now actors, and even investors, in the BMES due to political and renewable energy-based value formulae dictated from BMESs outside, for example the EU.

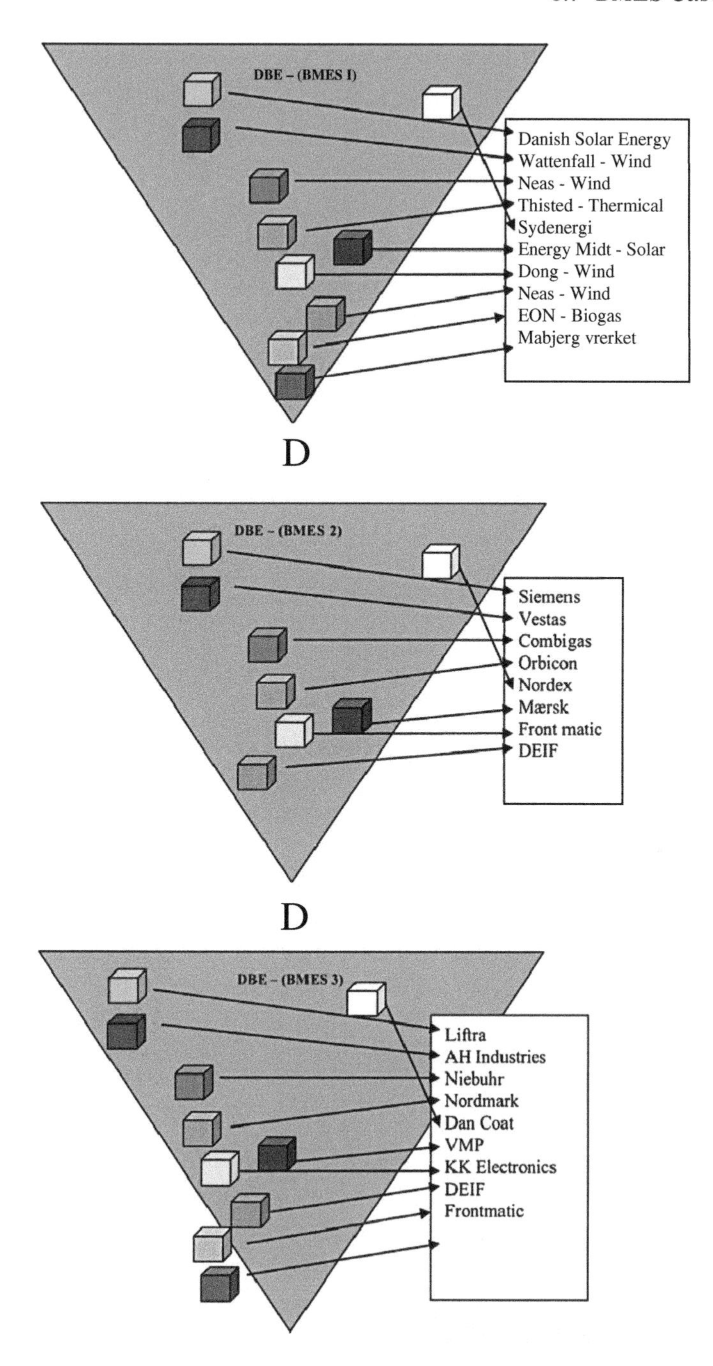

Figure 6.15 Vertical BMESs in Danish energy production.

6.7.3 Case 3 – Suppliers to Danish Energy Production BMES

The Danish Energy BMES has a tremendous number of suppliers in both Denmark and other European countries. Beneath we mention some of these different BMES seen in different contexts:

1. Oil BMES – Mærsk, Dong, Shell, Statoil, etc.
2. GAS BMES – Kosan, Praxair, EV Metalværk, etc.
3. Wind Mill BMES – Liftra, AH Industries, Nordmark, Siemens, Vestas, Niebuhr, KK Electronics, DEIF, DSV, etc.
4. Biogas BMES – Orbicon, Jenbacher, Gas2move, etc.
5. Solar BMES – Danish Solar Energy, Nordisk Solar, etc.
6. Termical Energy – Thisted Termical
7. Blue Energy – Foulum

Figures 6.15 and 6.16 show some elected vertical and horizontal BMESs.

6.7.4 Case 4 – HI – BMES to the Danish Energy BMES and Other BMESs

MCH is one of Scandinavia's largest and most flexible amusement centres with over 900,000 visitors each year. MCH has four BM portfolios – the Fair Center Herning, MCH Herning Kongrescenter, MCH Arena and Jyske Bank Boxen. MCH has the capacity to provide meetings for 15 people, conference space for 2,000 participants, football matches and arena space for 11,000 spectators and fairs for up to 50,000 guests. MCH's competence is to provide BMs and a BMES where amusements and business model exchange are core. Amusements can be a broad spectrum – rock, theatre, musicals and big sports events. MCH hosts and sets up more than 500 arrangements per year and is a market leader in setting up a BMES of amusement. MCH strengths are professional and service-minded employees, and up-to-the-minute facilities. Unique experiences and facilitating people and technology to meet each other are MCH's core competences.

MCH set up every second year an industry fair – a BMES – for the windmill industry and other industries from other BMESs. The industry fair, called HI Fair, functions as a BMES for five days. Many businesses with many different BMs operate in the HI BMES led by MCH, as can be seen in Figure 6.17. All BMs present at and under the HI BMES negotiate with MCH to be able to offer their BMs in the BMES.

Until now MCH has had very limited interest in relating to different BMESs but due to a decline in some of MCH's BMES they have decided to open up, for example, to the University BMES.

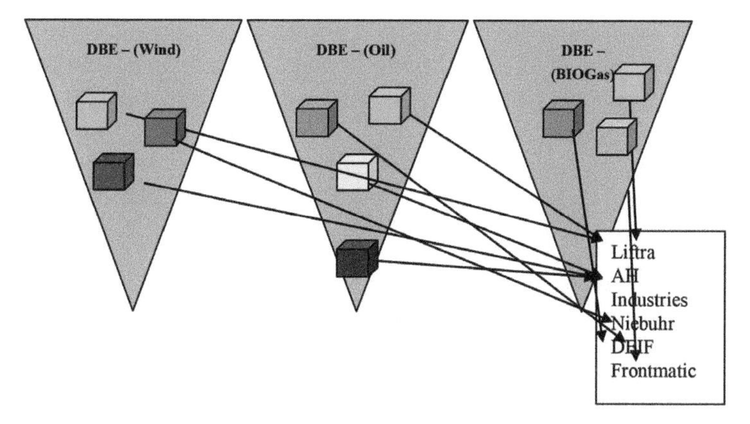

Figure 6.16 Horizontal BMESs in Danish energy production.

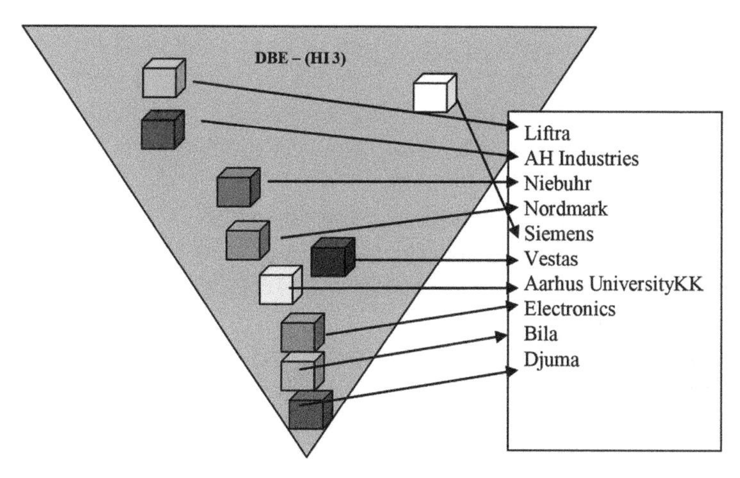

Figure 6.17 HI BMES set up by MCH.

6.8 Discussion

Today, most academics and practitioners consider the BM as a part of a market, industry, sector or a cluster – measurable, objective and one of a kind. Although there are many different definitions and types of business groups most define these related to a business model level but at a business level. We have earlier proposed that there is a need for a distinction between levels of business model focus, the business level and the business model level. We propose that the BMES core level should be focused on in research as "forming" an "umbrella" of "as-is" and "to-be" BMs represented in a specific BMES but also measured on related BMESs and BMESs that are not related – the BMES relation axiom. This is to prevent fuzziness and support discussion and further development of the BM theory.

Some BMESs together can form a group of BMESs that is interrelated – what we call a portfolio of BMESs – e.g. renewable energy BMES, fossil energy BMES – all focusing on energy production but measured in different viewpoints and contexts – either vertical or horizontally. These BMESs form a group of BMESs that have similarities due to, for example, the same customer focus, use of the same value chain, use of the same network, focus on the same mission – for example, energy production. Often the BMES portfolios like to be considered as interdependent, like Green Lab. Green Lab Skive is a business development park which will be located in a designated energy and resource landscape on the outskirts of Skive in the Central Denmark Region. The core of GreenLab Skive is a power-to-gas plant. GreenLab Skive is a symbiotic setup, where surplus energy and waste resources are used for testing, demo projects and other projects within green energy systems and green gas. Being part of the GreenLab Skive business development park will give you the opportunity to test your own technologies and projects in real time and within a full-scale renewable energy context (www.greenlabskive.dk). Sometimes each BMES in a portfolio competes with other BMES, sometimes they manage to "live" in symbiosis. As earlier mentioned, some BMESs, however, attract users who then attract customers to other BMESs in the BMES relations portfolio.

Further, we found businesses can be part of one (Vestas – Windmill) or even more BMESs (Siemens – Windmill, Hydropower, Solarpower). BMESs are where the business BMs operate and "exchange" their value proposition. The representation of BMs in different BMESs is a strategic choice of the business.

We propose that BMES business models and BMI should be viewed on different levels, as shown in Table 6.3.

BMESs can do BMES BMI at different BMES levels. The BMES vertical and horizontal level is considered as being complex but the BMES diversification is, however, the most complex level of BMES BMI – and is maybe therefore often not used by BMESs to secure their survival. The BMES Cube can be useful for downloading, seeing, sensing BMESs "on the way to begin operating" ("to-be" BMESs) and on BMESs "already operating" ("as-is" BMESs). It is possible to "innovate", "measure", "test", "download", "see" and "sense" any levels of a BMES. It is possible to "see" if the BMES can operate and how and why it is functioning or not functioning. It is possible to

Table 6.3 Levels of BMES

Levels of BMES	Characteristics of the BMES level
BMES component The smallest part of a BMES dimension	**BM's value proposition** components Value attitudes, attributes of different BMs **BM's customer and user** roles **BM's value chain functions** Primary functions: inbound logistics, operation, outbound logistics, marketing and sales, service Support functions: procurement, human resource management, administration, finance infrastructure, business model innovation **BM's competence** Product, production and process technologies HR – employees/people Organizational system Culture Network Physical, digital and virtual network **BM's relations** Tangible and intangible relations **BM's value formulae** **Profit and other value formulae**
BMES dimension	Value proposition Customer and/or user Value chain functions (Internal) Competence Network Relations Value formulae
BMES BMs	**BM of BMES** both "to-be" or "as-is" BM Cube
BMES BMES portfolio	**Group of BMs** that are interrelated in the BMES
BMES business	**Core business** level of a BMES with seven dimensions
BMES vertical	**BMESs** that are vertically linked together
BMES horizontal	**BMESs** that are horizontally linked together
BMES diversification	**BMESs** that are not linked together

see the BMES and its characteristics including dimensions and components at all different levels.

Summing up, we propose that any BMES consists of seven dimensions – six sides and the BMES relations inside the BMES that binds all the BMES BM's dimensions and components together and enables creation, capturing, delivering, receiving and consumption of values within the BMES.

6.9 Conclusion

There is until now not an accepted language developed for BMESs, nor is the term "BMES" generally accepted in the business model literature. This chapter commences the journey of building up a "language" on BMES based on case studies within the Danish Energy BMES, Suppliers to Danish Energy production BMES, The Danish Renewable Energy BMES and HI Fair BMES. The research shows that the old thinking of industry, sector and cluster systems defined these days is very much challenged because it gives the business and even the industry a kind of false security related to what really is the market, industry, sector or cluster. Especially when competitors or other business and BMESs begin to define the BMES differently – in a context-based way – then "conservative"-thinking businesses, industries and clusters are challenges; challenges because they lack strategies and competitive tools as many of them have formulated their strategy on the basis of market, industry, sector and cluster thinking – some would say old-school strategic thinking.

In contrast to the market, industry, sector and cluster definition we propose a different terminology – the business model ecosystem (BMES), defined as related to a context-based and viewpoint-based approach – including both "as-is" and "to-be" business BMs. We propose that any BMESs are defined in seven dimensions (value proposition, user and customers, value chain function, competence, network, relation and value formula). The BM is the focus as the smallest part of any BMES, contrasting with previous terms using the business as the focus. Each BM Cube can later be used to detail any BM in terms of dimensions and components (Lindgren and Rasmussen 2013).

The BMES framework and approach is built upon a comprehensive review of academic business and business model literature together with an analogy study of ecological ecosystems and ecosystem frameworks and studies of market, industry, sector and cluster terminologies.

The BMES today has to change fast related to the context or risk in the future of vanishing. BMESs may be considered to be established and look different from those we have seen in the past. A deeper understanding of BMES, seen in a context approach, could maybe give some answers as to why some BMES are successful and others not.

The chapter has addressed the concern with the difference between "the core business" of the BMES and the variety and strategy of its "as-is BMs" and "to-be BMs". If the distance between these becomes too large this can be a reason why the BMES falls apart or finds survival a challenge.

7

The Business Model Relations Axiom

Peter Lindgren and Ole Horn Rasmussen

Abstract

The notion of business models (BMs) has been used by strategy scholars to refer to "the logic of the business, the way it operates and how it creates value to its stakeholders" (Casadesus-Masanell and Ricart 2009). On the surface, this notion appears to be similar to that of a business model strategy. We present a conceptual framework to separate relations within any BMs and between any BMs in the business. BMs, we argue, are a reflection of business models' realized and unrealized relations. We find that in simple competitive situations there is a one-to-many tangible and intangible mapping between relations and the business BMs, which makes it difficult to separate the two notions. We show that the concepts of relations and BMs differ when there are important contingencies upon which a well-designed business model strategy must be based. Our framework delivers a clear separation between different relation viewpoints and BMs. This distinction is possible because we have verified through our research that relations are one of the seven dimensions of any BM and can be mapped internally to any BM and also externally – between any BMs internal and external to the business.

7.1 Introduction

The BM field has evolved substantially in the past 10 years. Business are now learning to "download", "see" and "sense" their business models – and from there their business models relations. Different approaches including business model canvas (Osterwalder 2011), the Stoff Model (Bouwman 2003, Bouwman et al. 2008), the open business model (Chesbrough 2007), the resource-based view (Wernerfelt 1984, 1995; Rumelt 1984; Penrose 1959), dynamic capabilities (Prahalad and Hamel 1990; Teece 1997) and game theory (Neumann 1928, 1944) have helped academics and practitioners understand the dynamics of business models and develop recommendations on how businesses should define their "as-is" and "to-be" business models.

However, drivers such as globalization, deregulation and technological change, just to mention a few, are profoundly changing the business model game and relations between BMs. Scholars and practitioners agree that the fastest-growing business in this new environment appear to have taken advantage of these structural business model changes to compete "differently" and innovate in their business models. Chesbrough (2007), Bower and Christensen (1995), Johnson et al. (2008), Markides and Charitou (2004), Casadesus-Masanell and Ricart (2010), Teece (2010) and Zott et al.'s (2011) studies show that businesses are actively seeking guidance on how to innovate their business models to improve their ability to create, capture, deliver, receive and consume value. One of the most important analyses in this work is the "downloading", "seeing" and "sensing" of their business models relations (Lindgren 2016a).

Advances in ICT have driven the possibility of mapping relations inside business models and business models' relations to other business models. Many businesses constitute "to-be" and "as-is" business models. Shafer et al. (2005) and Linder and Cantrell (2000) present 12 recent definitions of business models and 55 different business models (Gassmann et al. 2012) but hardly any mention of or focus on relations in business models.

Today practically all business models' tangible and intangible relations are possible to map and with more tools and evolvement of ICT it will soon be possible to have the full picture of any business models relations – both inside and outside BMs.

New relations for emerging business models steer researchers and practitioners towards a systematic study of relations to business models. Academics working in this area agree that for business to be effective and gain competitive advantage in different business model ecosystems in future, they need to develop novel relations inside their business models, between their business models inside their business, and between their business models and business models outside their business.

In fact, relation-based business model innovation that aims to reach the optimum of business multi business model innovation constitutes one of the most important sources of sustainability and growth of a business, but paradoxically – as far as we found it in our research – is also often neglected as a strategically important source or object in business model innovations.

Although it is relatively uncontroversial for business to innovate, managers of business model innovation must have a good understanding of how business models are related and how the BM's relations work. The academic community has – so far – only offered early insights on the issue. In truth, there is not yet agreement on what the distinctive features are of superior business

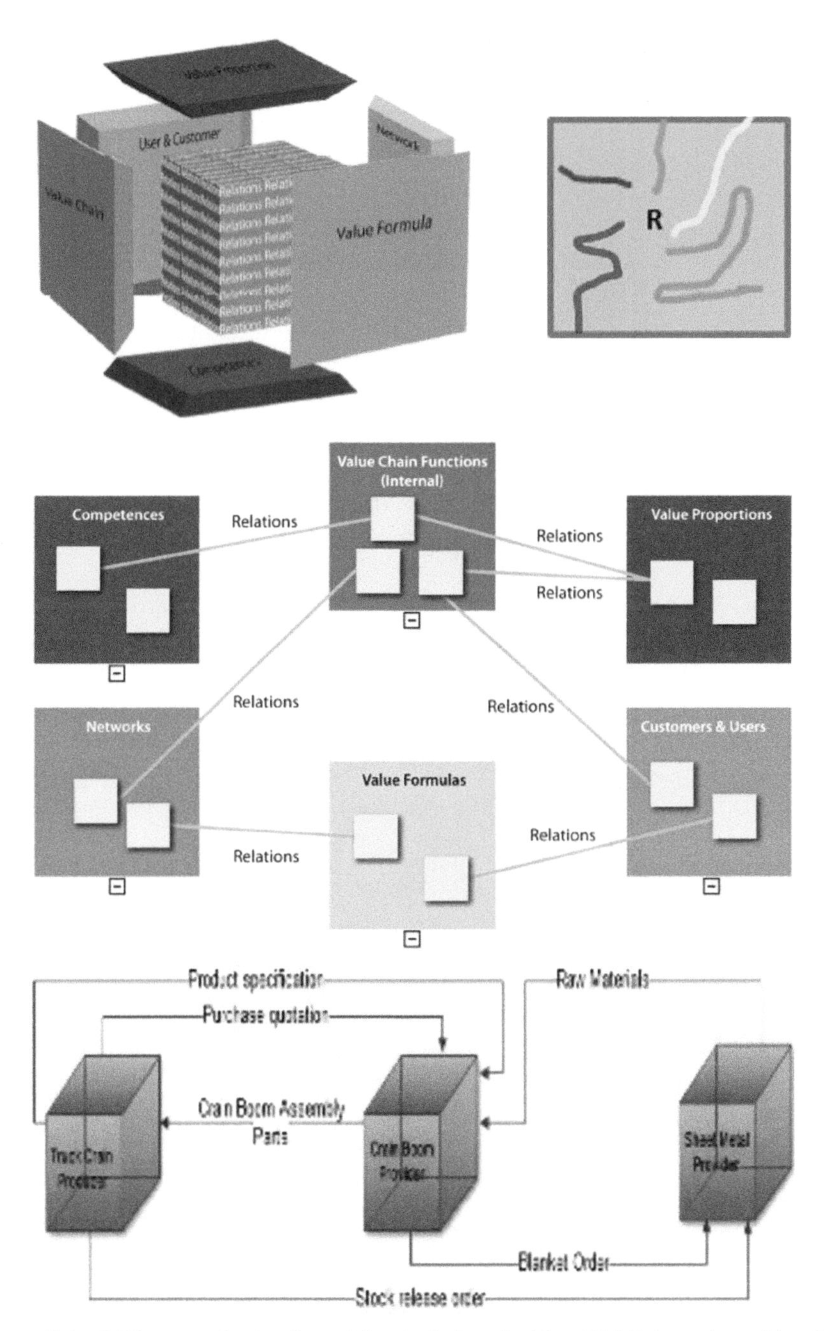

Figure 7.1 Different early sketch models of relations inside a BM Cube and outside a BM Cube (Lindgren and Rasmussen 2013, Lindgren 2016a).

models' relations and many academic business model frameworks even forget the relations of business models.

We believe that the dispute has arisen, in part, because of a lack of a clear distinction between the notions of business model dimensions, business models and business. The purpose of this chapter is to contribute to the research and literature by presenting an integrative framework to distinguish and relate the concepts of business model relations and business model relations' viewpoints.

As mentioned earlier in this book, business model refers to the logic and the framework we develop around the business, the BM Cube, the business model portfolio, the BM Cube's dimensions, the business models' dimensions' components (Lindgren and Rasmussen 2013) and the business present in the business model ecosystem (BMES) (Lindgren 2016b).

These relations enable the way business models operate and how the BM creates, captures, delivers, receives and consumes value for its business and its BMs.

"The way" in which one business model dimension is connected with another business model dimension is in our terminology defined as a relation. A relation relates all the BM's dimensions together and as can be seen in the sketch model of the BM (Figure 7.1) – the relations are placed firstly in the middle of the BM and then in a later approach between BMs.

In the sketch models it can be seen that relations are drawn, for example, as

- the relation of the value proposition to the customer dimension
- the relation of value chain function to the competence dimension
- the relation from one BM to another BM.

How can we get an overview of these relations and how can we use this overview to support operations of BMs and BMI in general? We will try and cover the answers to these questions in the paragraphs that follow.

7.2 BM Relations

Relations in our terminology consist of as "wire" and connectors. In Figure 7.2 a sketch model is drawn of how we hypothetically imagine what a relation looks like in a BM with a "wire" delivering and receiving connectors.

A relation works by relating BM dimensions and it enables BM dimensions to relate. In most cases, the relations relate BM dimensions' components from one BM dimension component to another. We identified (inspired by

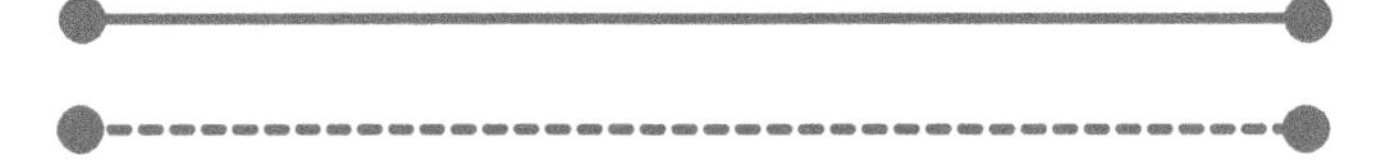

Figure 7.2 Sketch of tangible and intangible relations with a wire, delivering and receiving connectors.

Figure 7.3 No connection means no business and no "living" business model.

Allee 2008) very early two types of relations – tangible (unbroken line relations) and intangible (dotted line relations) relations, as seen in Figure 7.2.

Tangible relations are physical and visual – like a string, a wire, a pipe, a vein and/or an artery. **Intangible relations** are not visual, cannot be touched, are not "physically cabled". Examples are the wireless internet, mobile phone line, one person looking in the eyes of another, sound floating from one ear to another, smell floating in the air from a cheese to a person's nose.

Both types of relations are important in BMs – however, the physical and tangible relations have gained the largest attention and are most investigated.

Relations have to be connected to make them able to function. If a relation is not connected at both ends then values will not be able to flow from one BM dimension to another as illustrated in Figure 7.3. This can cause a lot of frustration but also leave businesses and business people in critical situations – even out of business (the Netflix case (Soteck 2014)). No wonder many business promote their brand as "being connected", staying connected and no wonder several politicians and grassroots groups are "fighting" to have connectivity as a "the right of the human being" (EU 2017).

Inspiration in our case and for the development of our BM relations terminology is as already comment on very much taken from Verna Allee's and Oliver Schwabe's framework (Allee 2008; Allee and Schwabe 2011) but also from the ICT and electrical power industry and healthcare scientific research.

7.2.1 Types of Business Model Relations

There are three kinds of relations that we found existing inside BMs and between BM dimensions. We found that these are always related from BM dimension components, either inside or outside a BM, to BM dimensions in BMs outside the BM. The BM relations can take different forms, which we give hereunder some examples of.

7.2.1.1 One-to-one relations

In a one-to-one relations, a BM dimension in Business Model A can have no more than one matching BM dimension component in Business Model B. A one-to-one relation is created if both of the related BM dimensions are primary keys or have unique constraints. The relations can be tangible or intangible as seen in Figure 7.4.

This kind of one-to-one relations we found is not so common in BM operation, because most information that is related in this manner would be either inside one BM or in very special BM operations between BMs. In this kind of relation, a BM dimension component in Business Model A can have one matching BM dimension in another BM dimension in Business Model A and/or have one matching BM dimension in Business Model B and/or Business Model C, D, and so on's BM dimensions. In Figure 7.4 we show one-to-one relations between two BMs.

Most often we found that there were many relations in a BM and its operation – often very complex to map and get an overview of. This can

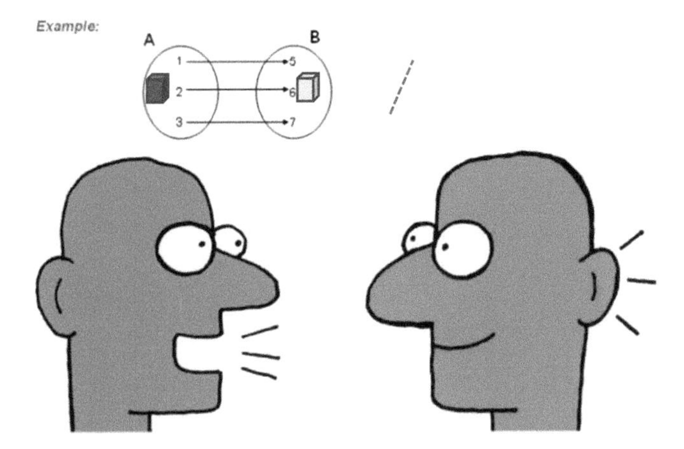

Figure 7.4 Example of one-to-one tangible and intangible relations.

be seen in the illustration in Figure 7.5 of two actual BMs in an insurance business.

In the insurance example in Figure 7.5 the customer BM dimension is "the BM dimension" in focus and is the BM dimension that most relations are connected to and values are transformed to and through. However, as can be seen in the example, the two BMs have very different set-ups from each other.

7.2.1.2 One-to-many relations

We found one-to-many relations very often in the BMs we studied. These kinds of BM relations are illustrated by an example in a sketch model in Figure 7.6.

In the model two BMs, A and B, have one-to-many to relations. Three BM dimensions in BM A each have two tangible BM dimensions in BM B. In this kind of relation, a BM dimension component in BM A can have many matching BM dimension components in another BM dimension in BM A and/or many matching BM dimensions in BM B and/or C, D, and so on's BM dimensions. That is naturally when it is considered that each business offers many BMs – the multi business model approach. In the example in Figure 7.6, Business A and its BM have a one-to-many relation to Business B's BM dimensions.

A one-to-many relationship is often created if one of the related businesses' BM dimension is a primary, core BM dimension or has a unique constraint. This can competitively be an advantage if the business can build many and different BMs on the basis of the one-to-many BM relations dimension, one BM relation to many BMs, one BM portfolio relation to many BMs. Several car and service businesses are experts on this business modelling type

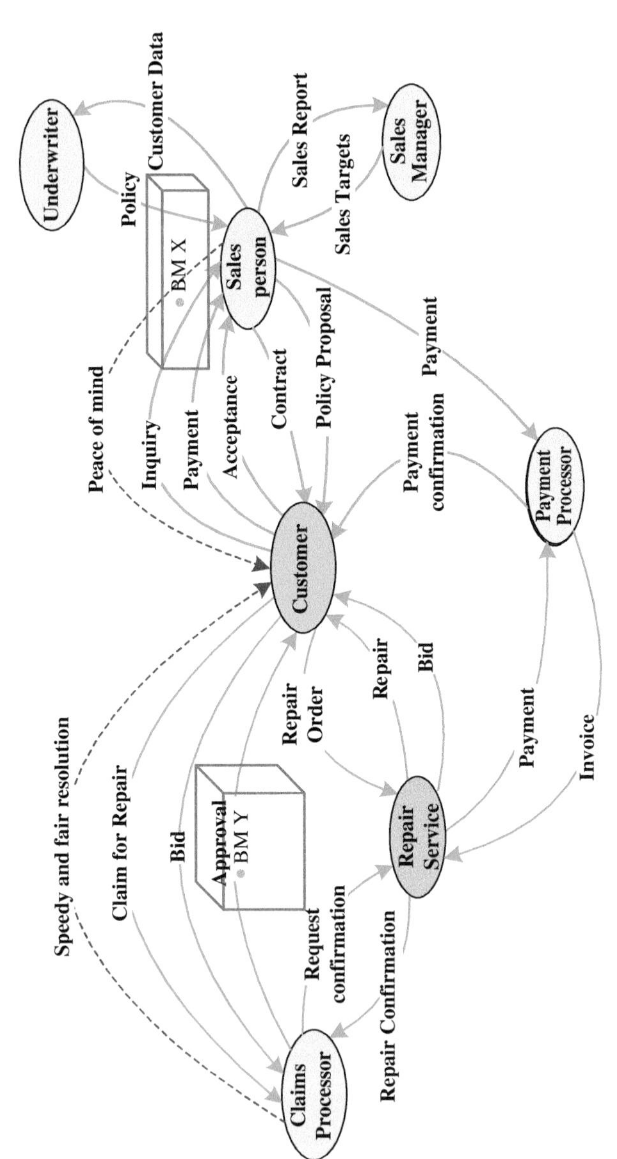

Figure 7.5 Example of BM relations in and between two BMs in an insurance business.

For example: One Department has many Employees.

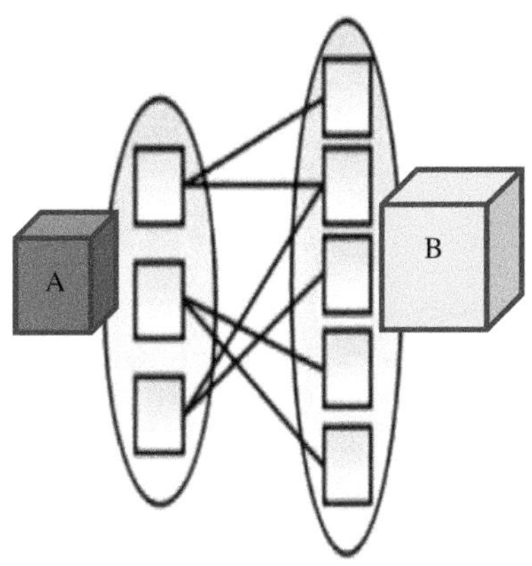

Figure 7.6 Sketch model of one-to-many tangible relations in a BM.

and this is often referred to as a platform strategy set-up. Here more BMs (cars, mobile phones, services, etc.) are built on the same production platform.

7.2.1.3 Many-to-one relations

In this BM relations perspective many BM dimensions or BM dimension components can be related to one BM dimension or one BM dimension component.

In Figure 7.7 two BMs, A and B, have a many-to-one relation. In this kind of relation several BM dimensions in BM A have one matching BM dimension

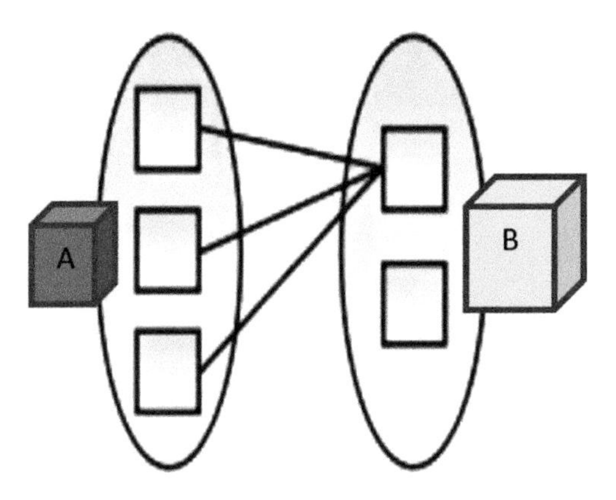

Figure 7.7 Sketch model of many-to-one BM relations.

in BM B and/or can have one matching BM dimension in BM B and/or BM C, D and so on's BM dimensions.

This knowledge can be used to optimize the BMs, to see "bottlenecks" and critical relations in BMs. This can hereby provide the "raw material" for later BMI.

7.2.1.4 Many-to-many relations

In a many-to-many BM relations terminology, a BM dimension in Business A (BM A) has many relations to BM dimensions in BM B – Quadrant 1 in the relations axiom. A BM dimension in BM A can also have many relations to BM dimensions in BM B, C, D etc. and vice versa in Business A – Quadrant 2 in the relations axiom. A BM dimension in BM A can further have many relations to BM dimensions in BM B, C, D etc., and vice versa in Business B, C, D etc. – Quadrant 3 in the relations axiom.

In Figure 7.8 we show two different examples of many-to-many BM relations. It can be seen that the relations perspective and mapping can become very complex (see the model at the right side of Figure 7.8). It is therefore necessary to have some structure and also some software to support the

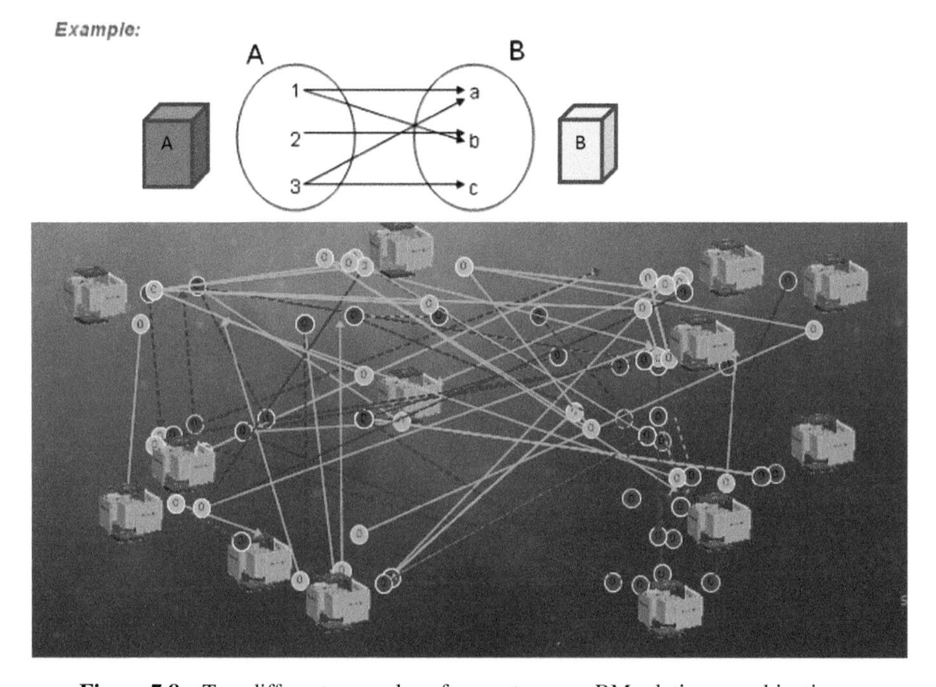

Figure 7.8 Two different examples of many to many BM relations combinations.

mapping and understanding of BM relations. We will later propose some tools to support this work.

7.2.2 Relations' Role

As argued and seen above, relations play a very important role in the process of business models' operation as they are the "arteries", "veins" and "nerves" that enable values to be delivered from one business model dimension to another. The relations are "the lifelines" between business model dimensions inside the BM, between BMs inside and outside the business. We found in our research that if the relations do not function then BMs will not be able to operate – and live.

Relations enable **the BM value process** of **creation, capturing, delivery, receiving and consuming values** (Lindgren 2016a) to take place. BM relations enable the BM value process of and by BM dimensions from the business model dimensions within business models and between business models – inside and outside the business – to take place. Without relations tangible and/or intangible BM dimensions – and values – cannot be transferred and BM dimensions and BMs cannot interact with each other. No business model therefore will be able to operate without relations – and we do no have, therefore, a "going BM".

Therefore we believe and also found in our research that there is a gap in BM research about BM relations. It seems to be a large mistake in BM research that there has not been more focus on BM relations and that relations in many BM frameworks and BM concepts are neglected and/or not included. The research lack is in BM relations, BM relations' role and how BMs transfer values to each other, how BMs communicate to each other and generally how they are able to create, capture, deliver, receive and consume values. In other words, the BM value process is not well understood in the BM community – especially related to the vital roles that relations play in BMs.

7.2.3 BM Relations Nodes, Hubs and Connectors

Fundamentally, what we have discovered is that relations must be connected in both the sending and the receiving "end". Our research shows very clearly that operating BMs have relations that have connections in "both ends". These connections – that connect BM dimensions – we call **BM relations node (BMrN)**. These BMrNs are called many things in businesses and academic literature. From a value network perspective they are called people or roles (Allee 2012) as seen in Figure 7.9; in technology based BMs and

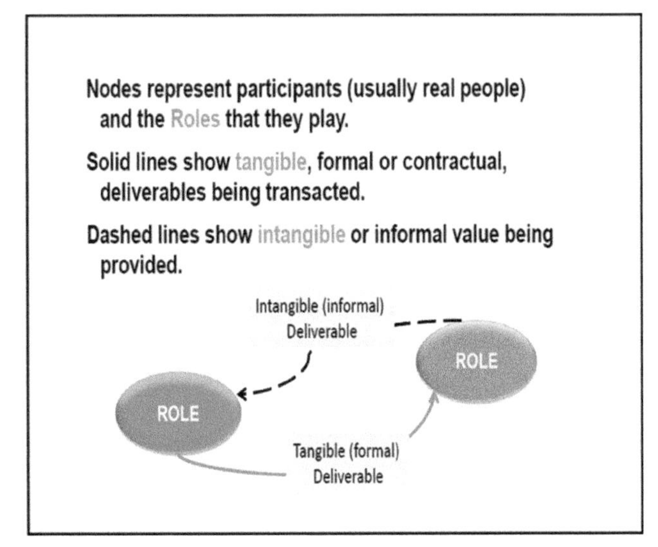

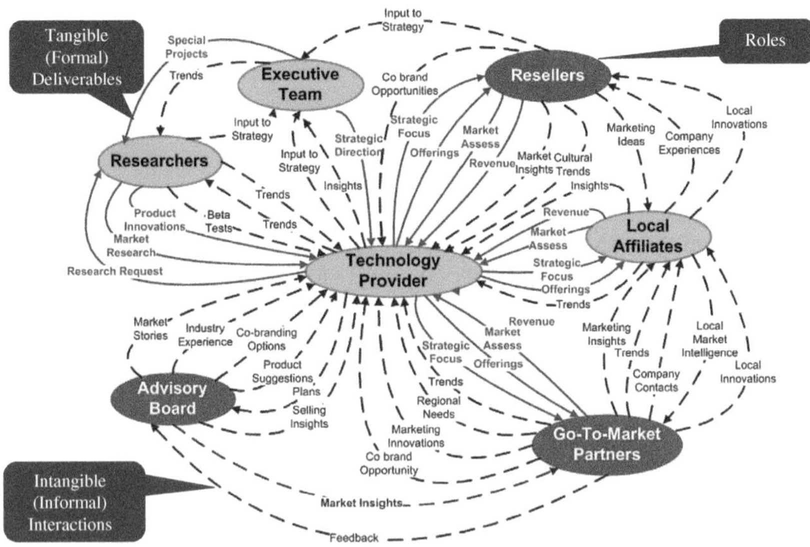

Figure 7.9 A value network with roles and deliverables (adapted from Allee 2012).

ICT systems they are called hubs; from the logistical perspective of storage centres, transport logistical centres and in post systems they are called mail boxes. Independent of what they are called, these nodes, hubs, connectors, etc. are extremely important in BMs as they enable the relations to deliver the value, "leave" the value from one BM dimension to be gathered by another BM dimensions relation. It also – if the hub function is correctly connected – secures that the value can be collected by another relation to be transferred to next destination – a BM dimension – in a receiving BM.

Nodes and/or a hubs or connectors – **BMrNs** – are therefore an important part for any business model. BMrNs are used to connect relations of one BMs or more – to enable BMs to operate "live".

We believe, as with ICT hubs, that a BMrN can contain multiple ports: **business model relation node ports** (**BMrNPs**). We have not tested and verified this yet in a multi case and sample perspective but several of our case studies show the presence of "multiple ports".

When a value arrives at a BMrN, it can be captured and transferred from the BMrN and to a BM relation to another BMrN of a BM dimension so that all business model dimensions or other business models can "see" or "receive" "a value" or "packet of value" through or at the BMrNP.

A passive BMrN serves simply as a conduit for the value, enabling it to go from one BM dimension to another. A so called intelligent BMrN, however, includes additional features that enable an administrator to monitor the "value packed" and the value process, passing through the BMrN. Further, an administrator can configure each BMrN. Intelligent BMrNs can also be classified as manageable BMrNs. These can – as we will see later in Part 2 of this book – be extremely valuable for the management and leadership of BM and BMI.

A BMrNP can actually be managed to read the destination address of each "value packet" and then forward the "value packet" to the correct BMrNP. These are extremely valuable as these BMrNPs can secure that every value is transferred automatically to the right BMrNP and BM dimension in a BM. In other words BMrNPs can have different competences – and they can be physical, digital or virtual.

Business models have both internal and external value interaction: internal value interaction between BM dimensions internal to the business model and external value interaction with business models within the business and outside the business. Business models have therefore a manifold of different relations through which the BM operates and can operate. The many different relations open to a business are by virtue of its business models what it employs in "as-is" BM relations and wants to employ in "to-be" BM relations.

In this context it would be more clear if it was possible to get an overview of the relations and find a generic structure to map the different relations a BM can have. This map should be able to plot the business models relations that are operating and those that are expected to be operating. We have proposed this mapping tool already (Lindgren 2012) and called it **the relations axiom** for "as-is" and "to-be" business models.

7.3 The Relations Axiom

To integrate the concepts of relations, business models and business, we introduced in 2012 an initial proposal on a generic four-square relations axiom framework for "as-is" BMs in a business, depicted in Figure 7.10.

The initial relations axiom was very valuable to get an overview of "to-be" BMs as it made it possible for the MBIT researchers to view BMs' relations from four different viewpoints and made us begin to study BM relations more deeply.

7.3.1 Quadrant 1 – the First Square of the Relationship Axiom – "in in BM Relations"

In **Quadrant 1** – **the first square of the relationship axiom** – we can study "**in in BM relations**". The process and focus are here to "download" and "see" the "relations of a BM's value creation, capturing, delivery, receiving and consumption" inside a business model inside a business.

As can be seen in Figure 7.11 the individual BM's dimensions are "bound" together and connected through relations "internally in the BM" – "the relation dimension". This view can be referred to as "the internal logic" and the "inside value transfer of a business model". It shows both the tangible and intangible value streams floating through the relations between the BM's dimensions. In this view it also shows the values transferred and value transfer inside the BM and inside the business. Relations going outside the BM cube are neglected in this "view" and therefore not shown.

Relations in this context can be divided into those that "deliver", "receive", "send" or "pass" value on to other BM dimensions and those who receive value from other BM dimensions. Relations are, as mentioned earlier, tied with BMrN to BM dimensions in the individual BMs' dimensions to be able to deliver and receive the "value package".

Relations that are not tied to a BMrN obviously cannot send, receive or leave values for "storage" or then later pass on values to a business model dimension. In the HSJD RPU use case shown in Appendix 9, we verify this statement and show BM relations that are not "tied" through a BMrN. These relations often end nowhere and the BM relation cannot therefore be of any use to the BM or the business. This is also important to note in the context of a BMI process or when "to-be" BMs are created. Often we observed that business created "to-be" BM relations but when these BMs were brought to operation they were never established and thereby never came to work. Also in this case those responsible for the BMI often forgot the relations or did not make an effort to connect the relations in the BM. This was very interesting for us to observe.

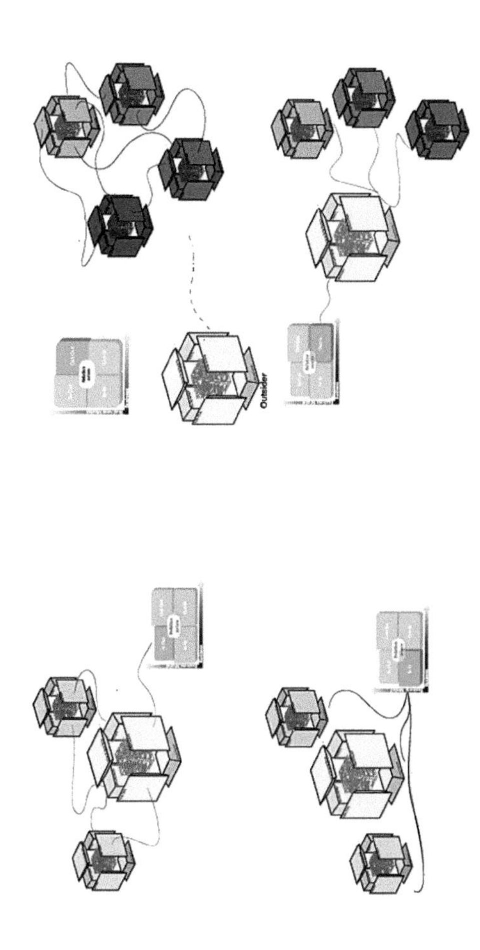

Figure 7.10 The relations axiom adapted from the first sketch model from Lindgren and Rasmussen 2012.

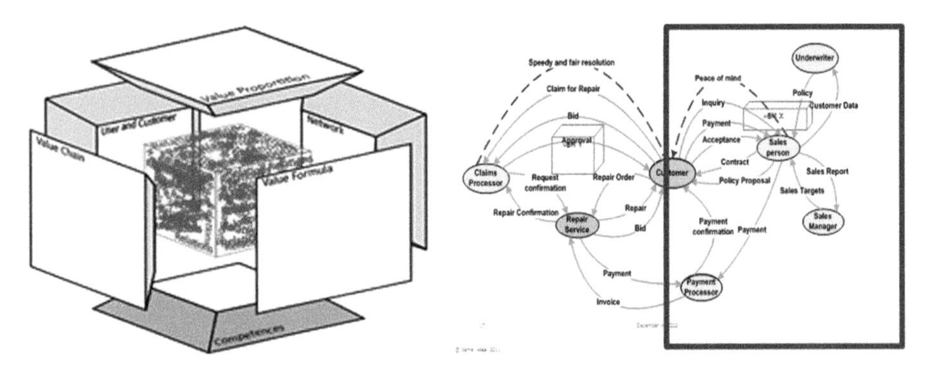

Figure 7.11 Quadrant 1 of the relation axiom – a sketch model of relations inside a BM.

Any set of business ("to-be") or ("as-is") BM relations are unique and different from one BM to another. However, we found that some BMs' relations can be copied to other BMs' relations and/or share other BM relations, and even quickly start to use, and perhaps take over, other BMs' relations. Therefore it is important to lead and manage BMs' relations as these relations could be considered as equal to arteries, veins and nerves in a body. Most businesses would like to have control of their BMs' relations – just as humans do not like others to have control of their arteries, veins and nerves – but not everybody can reject or refuse this as we will discuss later in Part 2 of the book, controlling and knowing the relations of BMs is very important (Amidon 2008; Allee 2012).

As we have already mentioned, BM relations consist of tangible and intangible relations that enable other BM dimensions to have value passing on and receiving tangible and intangible values. BMs and BMs' dimensions can be "stand alone" or "unused" relations only in the very early stage BMI phases. Otherwise they will slowly vanish and not be able to be used when needed, exactly like a hand or a leg of a body that has not been used for a long time – and then slowly withers. Relations must use and be related to other business model dimensions either inside or outside the business to enable the BM to operate and thereby become **"a going BM"** (SB 2009; William 2011).

In our research we also show that BMs are fundamentally network-based business models (NBBMs) (Lindgren et al. 2010) and most often connected through relations to other business models of users/customers and/or network partners. A business model that theoretically is isolated from other business models would vanish simply because it would not receive value and would not be able to create, capture, deliver and consume value to and from other BMs. The BM therefore would lose the "basis and purpose of life" – and would not be "a going BM".

Our research indicates that relations are perhaps one of the most important dimensions of a BM. Different domains of theory (intellectual capital, sales and marketing, global economy and others) also consider relations to be one of the most valuable parts of a business model and our economy (Håkansson and Snehota 1990; Amidon 2008; Allee and Schwabe 2011; Russell 2011). As Martha Russell (2011) said:

> Strategic value creation networks have become critically important in technology development and economic growth; co-creation relies on the relationship infrastructure of people, organizations and policies. These complex intangible relationship assets can be observed through network analysis of small, medium and large enterprises. By identifying relationships through which information and financial resources flow, visual insights toward a shared vision can be created and strategic network orchestration can be implemented. Using social network analysis, these relationship patterns can reveal competitive forces, gatekeepers and collaboration opportunities – within and across sectors – in internal and external innovation ecosystems around the world.

A BM that cannot be or is not related is worth nothing to others – or worth very little. It is therefore important to understand the relations between business models – and in this case both inside and outside the business – to get a full picture of the business models' logic, operation and potential.

Our hypotheses and proposed framework is therefore firstly that

> There exist relations inside any BM – the internal relations binding and connecting the business model dimensions in a BM together.

This is shown through two sketch models in Figure 7.12, Figure 7.13 and one sketch model from an empirically tested BM case, Figure 7.14.

As can be seen the empirical sketch model Figure 7.14 has only mapped the tangible BM relations (unbroken line relations). However, we know from our research that there were also intangible BM relations in this BM case.

We show and comment on this in detail in three different empirically verified business models inside three businesses' BMs (Vlastuin, Margit Gade and EV Metalværk) in Appendices 5, 10 and 11.

Our hypothesis is that each of the three different BMs shown in Figure 7.12, 7.13 and 7.14 are related in a unique internal relations spin and each of these are different to each other. This – we believe – makes any BM's relations unique and relevant to study individually and carefully. Every BM's "in in BM relations" in detail would potentially give us deeper understanding of the specific BM's relations construction and logic.

This relations mapping we can do related to "as-is" BMs – "downloading", "mapping" and in the "seeing" phase (Lindgren 2016a) – **the "in in**

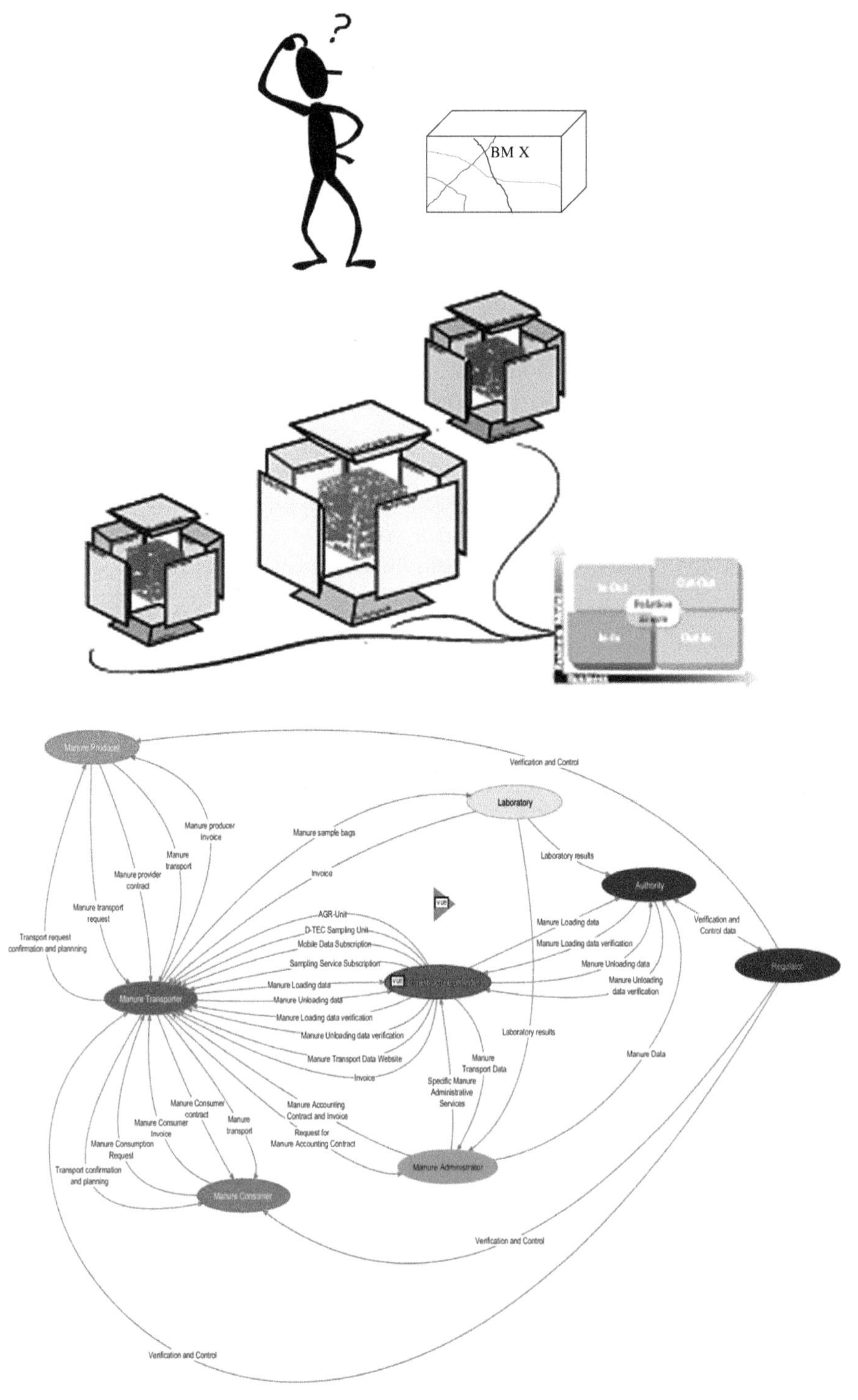

Figures 7.12, 7.13 and 7.14 Relations inside one BM; Relations in three different business models; Relations in an empirically tested BM case.

relations". We can also relate this to the "to-be" BMs – BMI phase – what we call "the sensing phase" (Lindgren 2016a). However, we propose that this work is separate from the "sensing phase" when businesses are creating the "to-be" BMs' relations. Therefore in Figure 7.15 we only show "as-is" BMs, but we will later (in Part 2 of the book) show "to-be" BMs' relations.

7.3.2 Quadrant 2 – the Second Square of the Relationship Axiom – "in out BM Relations"

The second assumption we made – and later empirically verified – related to the relations axiom we call "**in out relations**" (Figure 7.16). This was a proposal that claims that there exist relations from inside any BM and outside

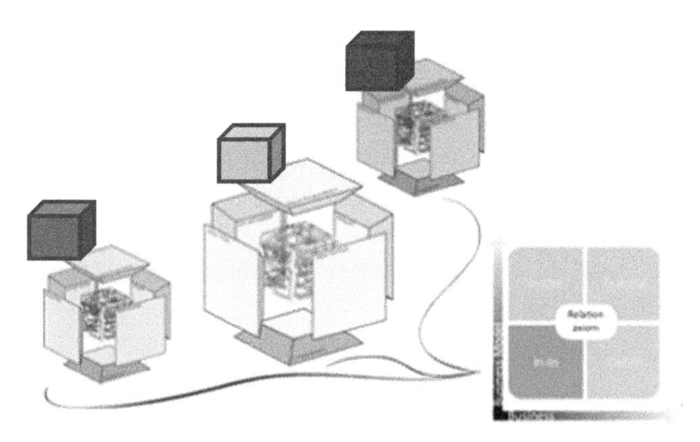

Figure 7.15 Three different relations inside three different "as-is" business models inside the business – "in in relations".

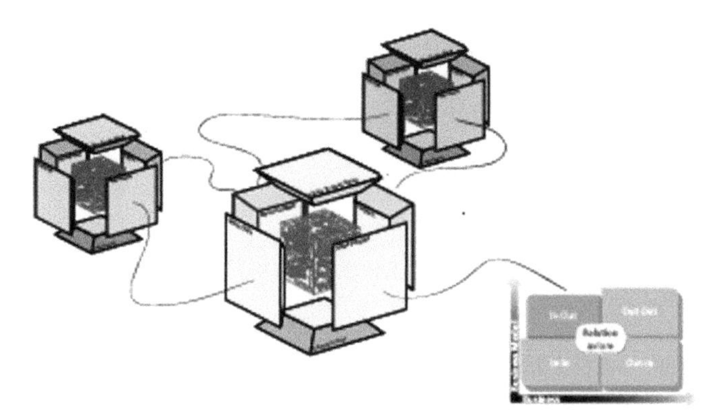

Figure 7.16 Quadrant 2 – relations inside the business and outside the business model "in out BM relations".

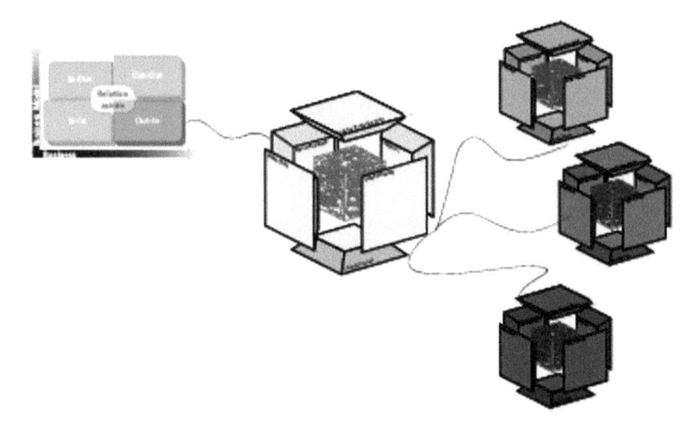

Figure 7.17 Relations outside the business to inside to the business model – "out in BM relations".

to other BMs. These relations – we propose – can be studied in two viewpoints or "modes". Firstly they can be studied through the relations between BMs inside the business but out of the BM in focus – quadrant 2 in Figure 7.16 – and secondly the relations between BMs outside the business and out of the BM in focus – quadrant 3 in Figure 7.17.

In Appendices 9 and 11 we show three examples of "in out relations" in two different businesses (HSJD, EV Metalværk).

7.3.3 Quadrant 3 – the Third Square of the Relationship Axiom – "out in BM Relations"

The third hypothesis we had – and we also verified in our research – was that there exist relations from inside the BM and outside to other BMs outside the business. These we call "**out in relations**" and are shown in Figure 7.17.

In Appendices 5 and 11 we show examples of "out in" relations in two different businesses (Vlastuin, EV Metalværk). We cannot tell yet if the relations are constructed the same as those operating inside the BMs but we have a hypothesis that they are. This we are testing in our MBIT Lab at the moment.

7.3.4 Quadrant 4 – the Fourth Square of the Relationship Axiom – "out out BM Relations"

The fourth hypothesis we had – and that we also verified in our research – was that there exist relations between BMs outside the business that the business model is not part of. These we call **the "out out BM relations"** and they are sketched out in Figure 7.18.

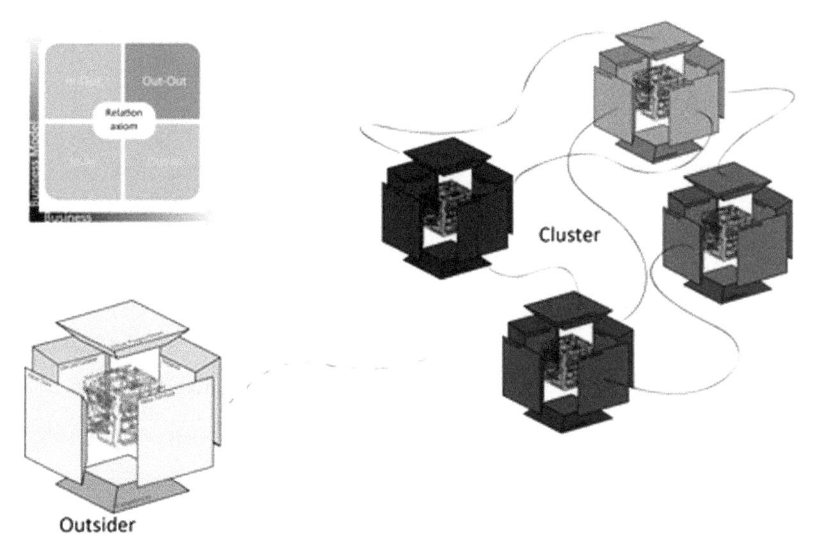

Figure 7.18 Relations outside the business model and outside the business – "out out BM relations".

In Appendix 11 we show an empirical case of "out out BM relations" in EV Metalværk.

By moving a BM from one relation axiom square – or **relation quadrant** – to another it is possible to see the BM in different BM relations perspectives – or **relation viewpoints**. This we found is highly valuable information not least in a later business model innovation context and process.

7.3.5 Relations with Different Characteristics, Functions and Contours

Summing up these hypotheses, tests and findings leads us to propose that there are relations inside any BM that have different characteristics, functions and contours. There are also relations connecting different BMs from outside the BMs and in this context both to BMs inside the business and to BMs outside the business. However we need more information and deeper research as we have only had limited time to study these BM relations. Further, we have not had time enough to study the cases sufficiently to stabilize our findings; neither have we had the opportunity to study the relations thoroughly in digitized BM environment. This we hope to be able to do in autumn 2017.

However our hypotheses are still that these different BM relations exist and it is not possible to explain BMs and BMI just through and by one set of relations as proposed by Granovetter (1973), Amidon (2008), Allee (2010) and Russell (2011). We believe that the BM relations are much more complex than proposed previously and that there lie several hidden possibilities in the "spaghetti of relations".

Our hypothesis is that a multitude of different relation types exist and it will be possible and valuable in the future "to see" these from different viewpoints to get the full picture of the business model relations, their characteristics and in what context they are operating.

In our research we found that BM theory until now has primarily focused on just a fragmented picture of BM relations (Osterwalder et al. 2005; Chesbrough et al. (2008); Johnson et al. 2008; Osterwalder and Pigneur 2010; Osterwalder 2011; Zott et al. 2011; Gassmann et al. 2012; Teece 2012) – primarily relations inside the BM and/or relations to customers and, to some extent, relations to suppliers.

We propose to increase this study by examining the types of relations that we believe exist following our preliminary research study, as outlined in Table 7.1.

Our hypothesis is also, therefore, that value deliverables passed through relations between network partners and network partners' BMs are many, much more complex and different to what we can and do see explained today in business model literature and even other academic work covering the theme of relations and value networks.

In the Table 7.1 we found some researchers that have commenced work on relations – often within a different perspective and with a different research approach to ours. However these works are opening up different dimensions of relations related to the BM environment and are highly valuable to our field of BM and BMI research. However, we believe that this work is not complete in understanding the relations in BMs and the creation and capturing of relations in BMI. They show us just some fragments of the BM's relation picture.

We therefore propose that relations related to BM and BMI should be studied much more and with more than just one set of relationship terminologies. We believe strongly it should be inspired and studied by and with an interdisciplinary approach and team.

In our relations axiom model we propose at least four sets of relation viewpoints of importance to BM study. This is also for the attention of BMI managers who relate to seeing the core challenges of BM relations.

7.4 Discussion

The approach in our research on the relations axiom was to use the relations axiom framework to "download", "see" and "sense" the tangible and intangible relations in BMs and to intellectual capital (IC) as sketch out in Figure 7.19

Table 7.1 Different types of relations related to BM and BMI

Type of relations	Frameworks available for inspiration and development	Examples/empirical cases
Internal relations in the BM	To some extent Verna Allee and Oliver Schwabe's framework (2011) can be used here – however, we consider that their work is only able to map parts of the relations in a BM. Their work is not complete in relation to the BM but more in relation to the value network	The insurance case; please also see the Neffics case (Neffics 2012, D 4.2) as seen in the right side of Figure 7.11
Relations between different BMs	To some extent Verna Allee and Oliver Schwabe's framework (2011) can be used here – however, we consider that their work is only able to map parts of the relations between BMs	Not available yet
Tangible and intangible relations between network partners' BMs	To some extent Granovetter's (1973), Håkansson's and Verna Allee and Oliver Schwabe's (2011) frameworks (D 4.2)	Not available yet
Tangible and intangible relations between network partners' businesses	Granovetter's (1973) and Martha Russell's frameworks; social networks (Davis 2009, Freemann 2009)	Available but in fragmented form
Relations between BMs' ecosystems	Martha Russell's framework can to some extent map relations between BM ecosystems but it is not especially focused on the BM ecoystem (BMES)	Not available yet

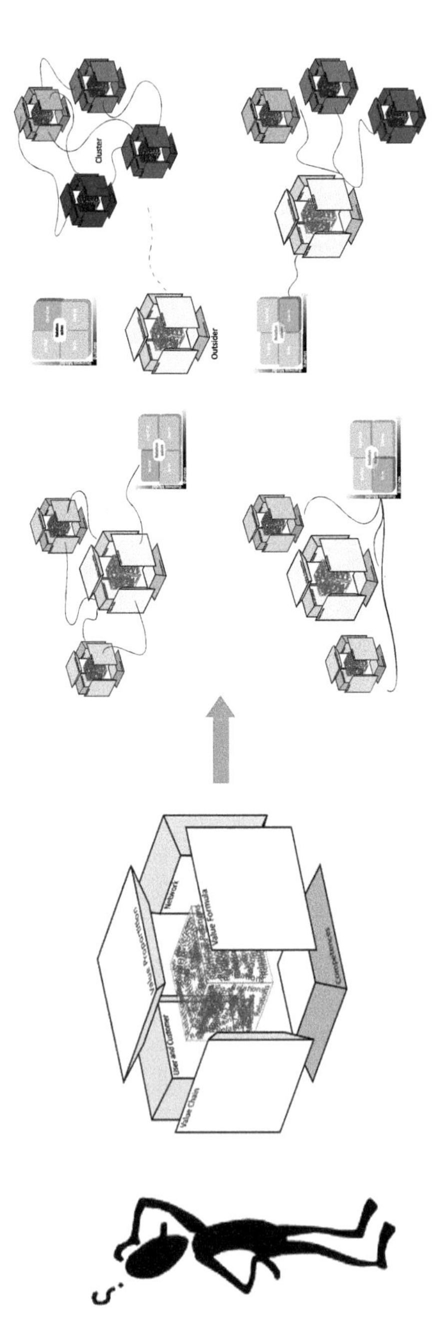

Figure 7.19 What is the BMI core challenge?

and explained in details in Table 7.1 in four business area viewpoint. In this process the businesses were encouraged to "download" their business "as-is" and "to-be" BMs' existing and potential relations to BMs and thereby IC. The business in question might not in their daily work be able to "see" and "sense" IC and potential relations to IC by themselves before mapping them in the relations axiom. Relations to BM competences and IC that in their daily business operations are valuable could now become visible and measured. When all relations for each BM were mapped – which we found was extremely time consuming – it showed "to-be" and "as-is" BM relations to competences and IC, which enabled the business potentially to understand better how many BMI possibilities and relations to competences and IC they really had.

In the mapping process two different types of relations-mapping approaches became valuable. Firstly the relations were mapped to each specific dimension in a BM – the BM Cube (Lindgren 2012). This work was done both for tangible and intangible values and relations. In our mappings above we have, however, only shown the tangible relations in Figures 7.12, 7.13 and 7.14. This was done related to creating, capturing, delivering, receiving and consumption of values sent through the relations. This means it is now possible for a business to "see" and "sense" which IC really has an impact and contributes to a certain BM and the BM's dimensions.

This work would have been much less time consuming and easier to do if it could have been digitized – or was digitized in the businesses. We argue that most of this information about relations to competences and IC are lying "sleeping" and "unused" already in the business ERP systems and as tacit knowledge within employees in the business.

The relations to competences and IC outside the business can also be quite simply found and mapped if customers and network partners are included in the mapping. For research purposes we used some supporting tools (Amidon 2008; Russell 2012; Allee and Schwabe 2011) to map the relations, which helped us to gain an overview of the value stream and relations between business BMs. More research, however, has to be carried out here to get a full picture as some information is lacking – especially "seeing" the relations and value stream from the outside (different BM innovation leadership viewpoints). This will be covered in Part 2 of this book.

In our research we found that the "out out" quadrant – the fourth quadrant – was very seldom used by businesses.

Methods of mapping intangible relations to competences, IC and hidden knowledge are vital to get the full relational picture of a business BM and relations axiom. Allee and Schwabe's value network tool (2008) is helpful to use when mapping intangible value and relations inside a BM and to some extent between BMs. Russell's (2012) and Amidon's (2008) relations tools are helpful when mapping relations more in a social network context and perspective – especially intangible relations to knowledge zones. However, more work has to be done to get a full picture, and MBIT researchers have already undertaken this work which we will comment on later in Part 2 of this book.

Some challenges related to the relations mapping tools used are still present. Mapping relations processes and mapping relations over time is especially a challenge. Allee's framework can, to some extent, help us to show us the value delivering and relations process – but there is still some development work to do to get a full "storyboard" of relations and transfer of IC over time. Russell's and Amidon's tools need also to be further developed to a kind of storyboard level taking the relations axiom to a mapping level that can show different times of relations.

Four cases each representing different businesses used different relations approaches to "download", "see" and "sense" relations. The "seeing" and "sensing" part were only done from the business viewpoint. The four businesses showed very different characteristics related the four quadrant in the relations axiom. These are detailed in Table 7.2.

When one analyses the characteristics and the relations axiom of the four BMI use cases individually it shows that the businesses are quite introvert in their work with their BMs – they use relations mostly with competence and IC internally to the business or relation to competences and the IC of close customers or network partners. It seems that there is much unused potential competence and IC for both "to-be" BMs and change of "as-is" BMs in the 1, 2 and 3 quadrants.

The cases studied point overall to a need to have more business focus on "downloading", "seeing" and "sensing" their relations on both their "to-be" and "as-is" BMs. Businesses have to "learn" their relations to IC and then learn how to release competences and IC strategically through their existing tangible and intangible relations. Businesses that try to release tangible and intangible competences and IC "blindly" often miss the real IC relation opportunities. They will not be able to "find" competences and IC that they are really looking for and which could create sustainable business model opportunities. Further, they might not even be able to release competences and IC which are vital for their BMs because they do not really "see", "sense" and understand the relations to competences and IC and interdependencies to relations of other BMs.

7.5 Practical Implications

Business can in the work of mapping relations to competences and IC benefit from using different methods to map relations. We propose that business has to

Table 7.2 Characteristics and relations to IC of different businesses

Cases	Characteristics	Quadrant 1	Quadrant 2	Quadrant 3	Quadrant 4
Vlastuin	Multi business model Primarily "in out"	Many tangible relations inside each business model to many competences	Many tangible relations between different business models inside the business	Many tangible relations between business models inside and business models outside the business	Several identified potential BMs and networks of BMs with unreleased tangible and intangible relations to IC
HSJD	Single business model "in out"	Many tangible and intangible relations inside each business model. Many tangible and intangible relations to many competences	Many intangible relations between different business models inside the business	Few tangible relations between business models inside and outside the business	Many identified tangible and intangible relations of BM and network of BM with potential unreleased IC
Margit	Single business model "in out"	Few tangible relations inside each business model. Many tangible relations to the same competence	Many tangible relations between different business models inside the business	Few tangible and intangible relations between business models inside and outside the business	Very few identified tangible and intangible relations to BMs and networks of BMs with potential unreleased tangible and intangible IC
EV Met-alværk	Single business model "in out"	Many tangible relations inside each business model. Many tangible relations to few key competences	Many tangible relations between different business models inside the business	Few tangible and intangible relations between business models inside and outside the business	Few identified BMs and networks of BM with potential unreleased tangible and intangible IC

work with different methods and hereby "learn" their relations to competences and IC. Six areas seem to be particularly important in this work:

1. Use user-friendly relations mapping tools for BM's relation mapping.
2. Use different relations mapping methods and tools.
3. Use methods and tools which can map business models' relations – both for "as-is" and "to-be" BMs.
4. Use BM relations methods and tools that can map value stream and relations over time.
5. Use different viewpoints related to mapping relations of BM (business model innovation leadership viewpoint).
6. Use methods and tools that can also show the implementation and operation part of value streams and relations – the "act–do" phase and part of relations of BMs.

Businesses these days have to get more knowledgeable about relations – both tangible and intangible – in their own business. They have to build up their ability to analyse and map structurally their relations in their BMs with the aim of innovating their business and BMs. Businesses today need to be more aware of their BMs' relations, which means they have to take enough time out to "download", "see" and "sense" their tangible and intangible relations both inside and outside their business. To do this they need to be able to "map" their relations, which has turned out to be very complicated and time consuming in some cases, as businesses often mix actual and perceived relations, finding it hard to keep these separated. We therefore propose to use the relations axiom to structure and guide this work.

In the process of mapping relations businesses also need beforehand to be aware of their potential relations and relations that they, or more precisely their BMs, are not part of – the "in out", "out in" and especially "out out" relations (the fourth quadrant). Mapping these is a question of "seeing" and "sensing" out of the box. This of course demands resources and time to go deeper inside and outside the business and its business models. In our research we observed more times that businesses often begin BMI without analysing carefully enough their BMs' relations and relations to IC. Thus they miss identifying where the business BMs' real and hidden relations to competences and IC really are and thereby find those relations to potential competences and IC that can be in many cases already used in their BMI.

The business can, when mapping relations to IC, face real revelations and new self-transcending knowledge about relations to competences and IC.

7.6 Conclusion

This chapter has shown the BM's relations axiom and the taxonomy of our proposal for a BM relations axiom. In the chapter we verify relations to competences and IC in BMs through "the lenses" and "viewpoints" of the relations axiom of business models. Hereby we show that it is possible to "see" and "sense" from the business viewpoint the business BM's tangible and intangible relations to competences and IC.

In the cases the businesses firstly mapped their relations by "downloading" their tangible and intangible relations of both their "as-is" and "to-be" BMs. The businesses then mapped them in a four-quadrant relations axiom:

1. **"In in" relations** – focusing on the relations of a business model to the other six BM dimensions. The viewpoint is from the single BM's side inside the business.
2. **"In out" relations** – focusing on relations of a BM to other BMs' dimensions inside the business. The viewpoint is from the single BM's side inside the business.
3. **"Out in" relations** focusing on the relations of a BM to other BMs' dimensions outside the business. The viewpoint is from the single BM's side inside out and outside in the business.
4. **"Out out" relations** focusing on relations of other BMs' dimensions outside the business which the single BM is not a part of or related to. The viewpoint is from the single BM's side outside the business.

Quadrant 1 and Quadrant 2 map relations to competences and IC inside the business whereas Quadrant 3 and 4 map relations to competences and IC outside the business. Different competences and IC release can be carried out and be expected to be carried out through the four different relations quadrant. Different quadrants and BMs "hide" different competence and IC potentials.

Mapping relations inside and outside BMs in a business is today very complicated and time consuming to carry out for managers responsible for BMI. There are today few tools that support BM relations mapping. Value network relations tools "tell" the business about value streams – both "tangible" and "intangible" – and social network relations tools "tell" the business about who is related to whom. When put into the relations axiom the competence and IC stream or potential value and IC transfer between BMs becomes visible. However, the tools still only show a fragmented picture of the relations axiom value and IC transfer and potential transfer – primarily Quadrant 1 and to some extent Quadrant 2 and Quadrant 3.

Conclusion: The Multi Business Model Approach

The BM Cube concept was evolved through our research which built on the increasing business model literature and practice in the early 2010s. The BM Cube concept initially came out of the research and test in the Neffics FP 7 EU project and was further developed and empirically tested in several businesses through the years 2010–2017.

Today, BM is argued in most academic literature to be a general model for how any business "runs" or should run its business – "a blueprint of the business". Conversely we argue that no business has just one BM – one model on which it runs all its business or intends to run its business. In other words the BM can be used for "as-is" and "to-be" businesses.

Our research points, however – in contrast to the other BM frameworks – strongly to the fact that businesses have more BMs – both "as-is" and "to-be" BMs – the multi business model approach. This was theoretically indicated already by Markides and Charitou in 2004 and in the Casadesus-Masanell and Ricart model in 2010 – but sadly nobody followed up on this in the BM community. It could have made a breakthrough in our understanding of BMs, BMI and strategic BMI.

Chapters 1–5 of this book have addressed the concern in the BM community and in BMI practice to just focus on the ideation and conceptualization of BMs. "BM canvassing" or innovating BM building blocks or BM dimensions when carrying out BMI – what we call "blind business model innovation" – is not sufficient to run a business. BMs and BMI – we propose – must address all different levels of BMI and all BMs in the business – both "to-be" and "as-is" BMs. All BMs are objects to BMI and should be to maximize the performance and sustainability of the business. The core business and all its levels of its business models – BM dimension components, BM dimensions, BM portfolio, business and BMES are all levels that can be objects for BMI – and should be considered for BMI.

The ICI and MBIT research addresses and documents a gap in BM research and the BM community – but also, however, a strong demand to find a generic definition and language of a BM (Teece 2010; Zott et al. 2011). The significance and importance of this work is related to the huge unexplored possibilities that we believe business model innovation offers today and can offer tomorrow. When we thoroughly understand the levels, dimensions and components of the business and its business models and are able to communicate, work and innovate with business models at all levels, then we will have achieved something – a next step in BM and BMI research and practice. Further, when the BM community agrees upon a common accepted BM language then it would be possible to achieve several advantages, as indicated in Table 3.1 in Chapter 3.

In this context, we proposed that any BM is related to seven dimensions — value proposition, user and/or customer, value chain functions (internal), competence, network, relations and value formulae. In the previous chapters we propose seven different levels of a BM from the most detailed level – the BM dimension component – to the BM dimension, BM, BM portfolio, business, and the vertical and horizontal business model ecosystem layer. The Vlastuin and HSJD case studies show examples of the BM Cube framework and the multi business model approach in practice and verify that the empirically documented seven dimensions and the levels really exist in the BM and BMI project.

Conceptually, the BM Cube was formed out of the seven dimensions and could be useful both in a 2D and a 3D version. The digitalization of the BM – as we will comment on and show later in this book's Part 2 – is highly necessary for us to achieve. When we are able to digitize the BM we will be able to take BMI and the understanding of BMs together with the business and BMES up a level.

The study has enlightened a strong demand for testing the BM Cube concept in a much larger business use case scale and sample than we have done until today – and especially in a digitized version. In the chapters that follow we show how we believe we can do this in the future and how we have prepared the research set-up to be developed further on the basis of wider quantitative and qualitative empirical research. To clarify more details of the BM Cube and see if the BM Cube also will function "in real life" a digitized version is our vision – and we believe this is a must. However this requires funding and businesses that, together with researchers, will be willing to participate and contribute. It also demands – we believe – that the research is carried out with the interdisciplinary approach as we have argued for in the chapters above.

Some funding for this MBIT research has been achieved in spring 2017 and the tests are now slowly being carried out as a part of a larger national-, EU- and US-funded collaborative research project. A network of established BM Cube labs spread out all over the world forms "the test bed". This we will comment on in this book's Part 2.

Appendices

Appendix 1: List of some of the Businesses Tested with the BM Cube Framework and the Seven Dimensions

Aarhus Airport
Aarhus Kommune
Aarhus University Library
AH Industries
Alvac A/S
Ardo
Arla
BHJ
Biogas2020 consortium
Brainbotics
Brande & Ikast Erhvervsråd
Calibeaut
Censec
Citylabs
Dansk Minkpapir
DEIF A/S
Den Magiske Fabrik
EV Metalværk A/S
Frontmatic
Genbyg Skive
Gobi Energy
GP Rådgivning
Green Lab Skive
Guldborgsund Kommune
Haldor Topsøe
Herning Municipality
HMN
HSJD
Human Company

Innovatoriet
IRD
KIC EIT Ship Demolishing
KMC
Kroma A/S
Lead
Løgesmose
Margit Gade
MCH
MesseC
Nettbusser
Nomi A/S
Praxair/Yarris
Salling Autoophug
Salling Entreprenør
SET, Herning
Skagen FF
Skive El Service
Skive Fjernvarme
Sotanæs Industrial Symbiosis
Strategy Reborn.Project: Stampe Elektronik
Subzidizer
Suntherm
Thise Mejeri, Veas, West Coast Lax
Trolhättan Biogas
Veng A/S
Vlastuin
X-FLEX

Appendix 2: Business Model Dimensions and Components

Table A2.1 Perspectives on business model dimensions and components, updating the work of Morris et al. 2003

Source	Specific components	Number	E-commerce/ general	Empirical support (Y/N)	Nature of data
Horowitz (1996)	Price, product, distribution, organizational characteristics, and technology	5	G	N	
Viscio and Pasternak (1996)	Global core, governance, business units, services, and linkages	5	G	N	
Timmers (1998)	Product/service/information flow architecture, business actors and roles, actor benefits, revenue sources, and marketing strategy	5	E	Y	Detailed case studies
Markides (1999)	Product innovation, customer relationship, infrastructure management, and financial aspects	4	G	N	
Donath (1999)	Customer understanding, marketing tactics, corporate governance, and intranet/extranet capabilities	5	E	N	
Gordijn et al. (2001)	Actors, market segments, value offering, value activity, stakeholder network, value interfaces, value ports, and value exchanges	8	E	N	
Linder and Cantrell (2001)	Pricing model, revenue model, channel model, commerce process model, Internet-enabled commerce relationship, organizational form, and value proposition	8	G	Y	70 interviews with CEOs
Chesbrough and Rosenbaum (2000)	Value proposition, target markets, internal value chain structure, cost structure and profit model, value network, and competitive strategy	6	G	Y	35 case studies

(Continued)

Table A2.1 (Continued)

Source	Specific components	Number	E-commerce/ general	Empirical support (Y/N)	Nature of data
Gartner (2003)	Market offering, competencies, core technology investments, and bottom line	4	E	N	Consulting clients
Hamel (2001)	Core strategy, strategic resources, value network, and customer interface	4	G	N	Consulting clients
Petrovic et al. (2001)	Value model, resource model, production model, customer relations model, revenue model, capital model, and market model	7	E	N	
Dubosson-Torbay et al. (2001)	Products, customer relationship, infrastructure and network of partners, and financial aspects	4	E	Y	Detailed case studies
Afuah and Tucci (2001)	Customer value, scope, price, revenue, connected activities, implementation, capabilities, and sustainability	8	E	N	
Weill and Vitale (2001)	Strategic objectives, value proposition, revenue sources, success factors, channels, core competencies, customer segments, and IT infrastructure	8	E	Y	Survey research
Applegate (2001)	Concept, capabilities, and value	3	G	N	
Amit and Zott (2001)	Transaction content, transaction structure, and transaction governance	4	E	Y	59 case studies
Alt and Zimmerman (2001)	Mission, structure, processes, revenues, legalities, and technology	6	E	N	Literature synthesis
Rayport and Jaworski (2001)	Value cluster, market space offering, resource system, and financial model	4	E	Y	100 cases
Betz (2002)	Resources, sales, profits, and capital	4	G	N	

Appendix 3: Monetary and Non-monetary Business Values

Man-Sze Li, Andrew Hinchley, Peter Lindgren, Jasper Lentjes, Henk de Man

We found that there were in principle many different ways to categorize business values, depending on the purpose at hand. In the Neffics treatment of value and in the BM Cube framework in general, we are guided by the notion of value flow between businesses' business models and specifically the addition of value in the course of business in which businesses and their business models are engaged. Very practically, businesses are interested in values in so far as they can be clearly identified, and are at least subject to some kind of measurement even if not directly measurable, in support of the pursuit of businesses vision, mission, objectives and strategies. In Table A3.1 we identified seven business value types defined and detailed more in the Neffics project work package 3.2 (Neffics 2012). Our proposition, based on the research carried out, is that business values that may be identified fall under these categories.

Business values and their business model values are created, captured, delivered, received and consumed through value inputs which, by definition, must be carried out by competences either by the BM itself, network competences or user or customer competences. For practical purposes, businesses are primarily interested in how a change in business values (identified through measurement) may relate to value inputs, leading to better value outputs as products, services and/or processes of products and services. Our study on tangibles and especially intangibles indicated above lead us to the following exploration of businesses' and business models' value types. Specifically, we map the business value types (as value inputs) to competences, as depicted in Table A3.2.

A3.1 Value Levels

A business or business model value can be variously described, interpreted, measured and used on different levels – the business model panorama view. Even if the value agents and the value activity involved are the same, there

Table A3.1 Tangible and intangible values

Tangibles	Intangibles
Financial	Negotiable form of intangible
	Resources
	Human capability
	Relationship
	Structural capability
	Regulatory

Table A3.2 Relationships between business value types and competences

	Tangibles	Intangibles		
STATIC ASSET / VALUE TYPES	Financial assets	Human competence	Internal structure	Business relationships
Financial improvement, being directly measurable	Yes			
A change in a negotiable form of intangible			Yes	Yes
A change in resources			Yes	
A change in human capability		Yes		
A change in relationship assets				Yes
A change in structural capability			Yes	
A change in regulatory requirement	May be	May be	May be	May be

could be different outcomes and interpretation of the value activity process depending on the level at which the value activity process is analysed. There is today no definitive value analysis that applies to all business cases and business model contexts.

Neffics introduces the concept of value levels, as shown in Table A3.3.

It is a critical requirement that value flow should be viewed in its entirety across a network of business models stretching across multiple business models and businesses. This is a major separation from many earlier approaches, particularly value chain methodologies where there were constraints both on the types of value involved and on the limits of the chain itself. Sveiby states (Sveiby 2001):

> In contrast to the Value Chain methodologies argued that the intangible value in a Value Network – network of business models – grows each time a transfer takes place because knowledge does not physically leave the creator as a consequence of a transfer. As Value is contextual and a knowledge transfer can also be viewed as a loss. So calculating the actual value of the knowledge transfer more accurately requires being able to understand the value impact of each unique transfer.

Hereby each business model is involved or related.

Table A3.3 Value levels of business value

LEVEL	DESCRIPTION
L1	**Value at the level of the person and/or things** This correlates to a person and/or things involved directly in a value activity (typically an employer, employee, user, customer or a network partner). *Note: it is possible, and may indeed be required, to discuss value at the level of "things" also, which are the other main categories of entities in our value landscape. The argument is that machines become more "intelligent", possess a myriad of self-properties, and have an increasingly prominent role in a value activity which needs to be separately accounted for as entities "in their own right", independent of the processes in which they are engaged.*
L2	**Value at the level of the individual process** This correlates to the activity of a business as reflected in or captured by a process within the business.
L3	**Value at the level of the individual business unit** This correlates to a discrete component of the business contributing to the overall business (whole). It typically comprises people, things and processes.
L4	**Value at the level of the individual business (typically a business)** This correlates to the business as an integral whole for pursuing a range of activities to meet a specific goal; it is typically (defined) within the business model of a business and is expressed through the value proposition of a particular business (value creation, capturing, delivering, receiving, consumption).
L5	**Value at the level of the business model ecosystem** This is correlated to the business model ecosystem within which a business operates with its BMs and is expressed through the value of an offering (one or more value objects) to the business model ecosystem for the offering encompassing both its supply and demand.
L6	**Value at the level of the whole business model ecosystem** This is correlated to the business model ecosystem as a whole where the focus is on the value of an offering relative to those of other offerings in the same or other related business model ecosystems, as part of a larger "value package" that drives supply and meets demand in aggregates.
L7	**Value at the societal level** This is correlated to the broader societal value as determined by culture and social norms and other drivers.

These value-related activities are modelled in this document as flows where changes in value occur. By adapting Sveiby's work, we arrive at a value flow framework in positioning business value categories in relation to business network complexity.

A3.2 The Economics and Measurement of Business Values

Economics as a discipline arose from the need to allocate resources: the resources available are limited – either by nature (as in natural resources) and/or by infrastructure (as in the power plant for energy and the network infrastructure for telecommunications). In economics, a distinction is generally made between supply and demand. For the purpose of this chapter, what is of most interest is the contribution of economic theories and models to the flow of value between supply (the value sender) and demand (the value recipient), and how a change in value may be accounted for in the value flow. This helps determine what to measure and how to measure it in business value and business model value analysis.

The FP7/ICT COIN Project has provided detailed analysis of the following economic theories and models in assessing ICT-based services for enterprises in relation to their utility and value-added properties (COIN 2011):

- The efficiency model: efficiency-driven competition and competitive markets to produce the "best" price through the market as an "invisible hand"
- The value chain model: "margin" as the competitive differentiator and competitive strategy
- The transaction costs model: transaction as an organizational concept and cost as an economic friction
- The resource-based view: competency as the key resource and the application of a bundle of heterogeneous and not perfectly mobile resources for creating sustainable competitive advantage
- The game theory model: decision support to prevent lock-in situations; arguments for commonly defined solutions
- The coordination model: coordination of structures, resources and people for advancing economic welfare
- The network economics model: positive feedback and critical mass as features of network economy; new notions of network externalities and demand-side economies of scale
- The new institutional economics model: "institutions matter" – broadening the scope of economics from resource allocation to the broad context for institutional arrangements
- The innovation economics model: assessment of innovation uncertainty and success; argument for innovation as a growth driver.

Table A3.4 The value types addressed by economic models

VALUE TYPE	ECONOMIC MODEL								
	efficiency	value chain	transaction costs	resource-based	game theory	coordination	network economics	new institutional economics	innovation economics
Financial improvement being directly measurable	✓	✓	✓						
A change in a negotiable form of intangible								✓	✓
A change in resources	✓	✓	✓	✓					
A change in human capability				✓					
A change in relationship assets			✓	✓	✓		✓		
A change in structural capability				✓	✓	✓	✓	✓	✓
A change in regulatory requirement								✓	✓

Our research indicates that these economic models have a focus on different value types.

Moreover, the economic models address values and value flow at different value levels and types as indentified and described in Tables A3.3 and A3.4. A summary is given in Table A3.5.

From our assessment, it would seem useful to apply the economic models to value flow analysis in respect of the different value types and on the different value levels to which the economic models are individually mapped. This could eventually yield a set of specific business values and value measures that would, potentially, contribute to a better understanding of value analysis, within the rigours of the economic science. Econometric techniques may be used to devise value measurement mechanisms.

Sveiby gave some methods for measuring intangibles. Sveiby (2010) categorizes these into the following four categories. Sveiby states his categorization is an extension of the classifications suggested by Luthy (1998) and Williams (2000).

- **Direct intellectual capital methods (DIC).** Estimate the $-value of intangible assets by identifying their various components. Once these components are identified, they can be directly evaluated, either individually or as an aggregated coefficient.
- **Market capitalization methods (MCM).** Calculate the difference between a company's market capitalization and its stockholders' equity as the value of its intellectual capital or intangible assets.
- **Return on assets methods (ROA).** Average pre-tax earnings of a company for a period of time are divided by the average tangible assets of the company. The result is a company ROA that is then compared with its industry average. The difference is multiplied by the company's average tangible assets to calculate an average annual earning from the intangibles. Dividing the above-average earnings by the company's average cost of capital or an interest rate, one can derive an estimate of the value of its intangible assets or intellectual capital.
- **Scorecard methods (SC).** The various components of intangible assets or intellectual capital are identified and indicators and indices are generated and reported in scorecards or as graphs. SC methods are similar to DIS methods, except that no estimate is made of the $-value of the intangible assets. A composite index may or may not be produced.

Sveiby (2010) has also graphically represented all known approaches (Figure A3.1), together with a complete listing.

Table A3.5 The value levels of economic models

VALUE LEVEL	ECONOMIC MODEL								
	efficiency	value chain	transaction costs	resource-based	game theory	coordination	network economics	new institutional economics	innovation economics
L1 (Person)				✓	✓	✓			✓
L2 (Process)		✓							
L3 (Business unit)		✓	✓	✓		✓			
L4 (Business/firm)	✓	✓	✓	✓		✓	✓	✓	✓
L5 (Market)	✓	✓	✓	✓			✓	✓	✓
L6 (Ecosystem)							✓	✓	✓
L7 (Societal)								✓	✓

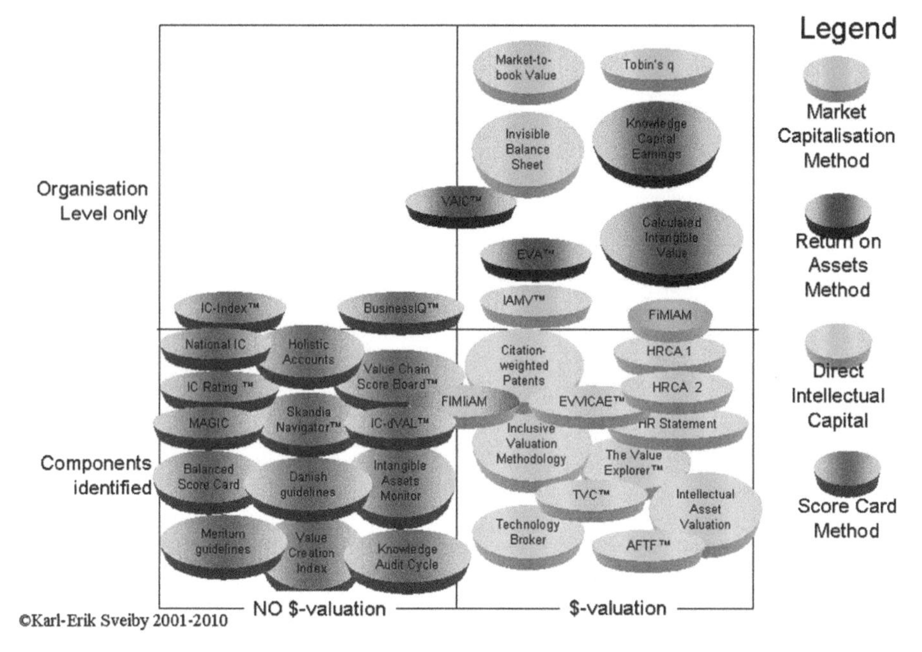

Figure A3.1 Intangible assets measuring models (Sveiby 2010).

The balanced scorecard is important to include in this discussion as it was used in 2013 in more than 50 per cent of European and Asian businesses and 50 per cent of US larger businesses. Balanced scorecard uses a very limited number of measures in each of a number of (independently assessed) sub-parts of an organization's activity. Balanced scorecard tends to focus on general objectives in a particular area rather than promoting specific measures. Its key core financial measures of return on investment (ROI), profitability, revenue growth/mix and cost reduction productivity may equate to different formulae in different contexts. Similarly balanced scorecard cites market share, customer acquisition, customer retention, customer profitability and customer satisfaction as important measures, but leaves the detail to be worked out closer to the user.

Please see Neffics D 3.2. www.Neffics.eu.

Appendix 4: Strategic Value Map: Blue Ocean Strategy Framework

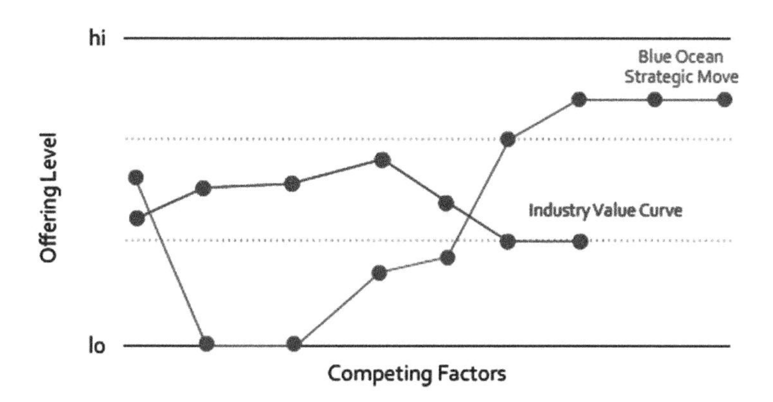

The strategy canvas is the central diagnostic and action framework for building a compelling blue ocean strategy. The horizontal axis captures the range of factors that the industry competes on and invests in, and the vertical axis captures the offering level that buyers receive across all these key competing factors.

Figure A4.1 The strategic value map from the Blue Ocean Strategy framework (Kim and Mauborgne 2005).

Appendix 5: Vlastuin Use Case

Vlastuin located in Netherlands, started operating in 1959. Vlastuin employs around 150 people and had a turnover of 27 million euros in 2011. During its more than 50 years, Vlastuin has added more BMs to its business and thereby slowly increased its core business. It started off by installing and servicing furnaces and boilers, gradually developed manufacturing and later on added the assembling of cranes and parts to the business. A graphical representation of Vlastuin's business evolution can be seen in Figure 4.14 in Chapter 4.

A5.1 Business Case 1: Vlastuin's Crane Business

The first business case provided by Vlastuin is production of the crane booms. This business started due to the evolution of the crane producers (customers) outsourcing crane boom production (value chain functions). A crane boom is the extendable and retraceable arm of the crane (product) which lifts the loads. See Figure A5.1.

Vlastuin as a manufacturer of D-Tec container trailers had **competences** (**C**) of accurate bending and high quality welding (**production and process technology** and **HR**) of large heavy pieces of steel, which was exactly what crane producers were looking for. Currently, Vlastuin is a provider of the crane booms (**value proposition** – **VP**) to crane manufacturers throughout Europe. The truck crane BM involves three major stakeholders: truck crane producers (**customers** – **CU**), crane boom providers (**network** partner – **N**) and metal sheet suppliers (**N**). Each of these will be shortly introduced presenting their roles and interconnections between each other.

A5.1.1 Truck crane producer business case (OEM customer)

Truck crane producers, as the name implies, produce the cranes and mount them on the truck. Often they outsource part manufacturing and focus more on the final product. Part of the outsourced manufacturing is boom production in which Vlastuin specializes. The truck crane producer has extensive knowledge (**C**) on crane boom manufacturing since it was originally manufactured in-house. Therefore, it demands the same or even higher quality for the outsourced parts (**VP**). Furthermore, in this specific crane boom part provided by Vlastuin, the truck crane producer also has a contract with a metal sheet supplier to ensure that the raw material meets the specifications for manufacturing (**VP**).

Figure A5.1 An example of a Vlastuin crane boom on trucks.

A5.1.2 Crane boom provider (Vlastuin)

The crane boom provider, or in this case Vlastuin, manufactures (**value chain** – **VC**) crane boom parts based on customer specifications (**VP**). This process starts with the creation of the production drawings and product quality plan (**VP**) by a specialized engineer. Afterwards, special sheet metal is ordered from the supplier (**VC**). After raw materials are received the production processes launches (**VC, C**). Three major steps in production are laser cutting, sheet bending and certified welding (**VC**). Laser cutting involves cutting out various boom components of the sheet metal plates using a laser. This provides high quality cutting edges and very precise component dimensions. Sheet bending is where high dimension heavy components are bent at right angles according to predefined sequences. In order to obtain exact bend angles, very precise laser angle measurements are performed during the process. Certified welding is performed with high-end welding equipment by certified welders (**C**) due to safety regulations of truck cranes. Here, the separate boom components are welded together in a pre-set welding order. This is to avoid the crane boom getting twisted due to the heat transfer and thick metal, causing problems later in crane boom operation. After all the production processes are carried out and quality is ensured, separate welding assemblies are grouped together and sent to the customer production line (**VC**).

Below, we have summarized the value chain function and process that Vlastuin addresses. It also indicates some of the tangible and intangible value propositions that Vlastuin takes care of together with some of the competences embedded in Vlastuin's BM. Further, it gives an overall view of the **relations** (**R**) inside the specific BM.

A5.1.3 Sheet metal provider

Specification-meeting sheet metal is supplied by a sheet metal provider after the truck crane provider sends out a stock release order assigning certain amount of stock to the crane boom provider. Due to its long manufacturing processes this is manufactured in batches and kept in stock. After receiving an order the sheet metal is transported to the crane boom provider.

For an overall graphical overview of the Vlastuin crane business case, we have drawn up three BMs in action with Vlastuin BM at the centre in the Figure A5.2.

One building block is not shown. Our comments regarding the value formula of the crane boom provider Vlastuin is confidential information. In the next case we will, however, be able to go a little deeper into another of Vlastuin's BMs.

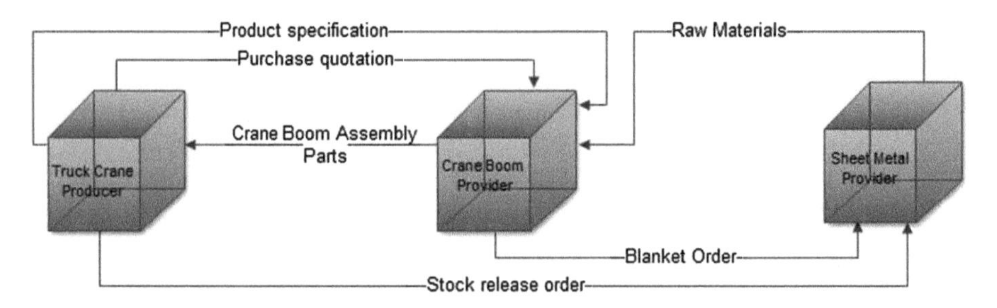

Figure A5.2 Vlastuin cranes business case overview.

A5.2 Business Case 2: Vlastuin's Paperless Manure Transportation Business Case

Vlastuin is also in the manure transportation data administration business. In the Netherlands, it is decided by law that in order to transport manure, authorities have to be notified at the start and at the end of the transportation with manure samples. Due to these regulations, Vlastuin started providing AGR units (Dutch for Automatic Data Registration) (**VP**). This unit sends data (**VP, VC, C, R**) to the Vlastuin server where it is filtered and forwarded to the authorities (**user**). By doing this, it dramatically decreases the processing time and paperwork needed for manure transportation (**VP**) for the user and customers (**CU**). There are eight significant stakeholders in this business case, which will be introduced next.

A5.2.1 Manure producer

A manure producer is usually a livestock farmer (**CU**'s customer) who has excessive amount of manure. Farmers usually have a contract with the manure transporter (see A5.2.3) (**CU**) which means that all the work that comes with manure transportation is done by the manure transporter. Some examples could be that the manure transporter is responsible for finding manure consumers (**CU**'s customer), or the manure transporter is responsible for all the paperwork around the manure transportation (customers' **value proposition** demand). The cost associated with manure transportation is deducted from manure producers' payment for manure. The manure producer gets a digital version of the paperwork from the manure transporter.

A5.2.2 Manure consumer

The manure consumer (**CU**'s customer) is usually the farmer who needs the manure as fertilizer for his or her fields (**CU**'s customers' (upstream) **value proposition** requirement). The manure consumer has a contract with the manure transporter which includes all the work associated with manure transportation (**CU**'s customers' (downstream) **value proposition**). Manure consumers get the invoice for manure together with the digital copy of the paperwork.

A5.2.3 Manure transporter

Manure transporter is the direct customer (**CU**) of Vlastuin. This usually is the transportation company which transports manure from manure producer to manure consumer. The manure transporter has a contract with both manure producer and consumer, and dispatches tank trailers to manure producers upon request. During loading of manure to the tank, samples of the manure are packaged into the sealed bags, as can be seen in Figure A5.3.

These samples are fitted with barcodes (added value proposition) which are scanned and sent to the authorities together with other required information (**VP, VC, C, R**). This is automatically performed by the AGR unit via an infrastructure provider service (internal business network partner (**N**) value proposition). After receiving confirmation from the authorities (**N**) about successful transmission, the manure is transported to the manure consumer (**CU**'s customer). The manure consumer is automatically determined by GPS data (added value proposition) combined with manure administration data (added value proposition) thus identifying the closest manure consumer location. Before transportation, the consumer will need to confirm if he or she wants to receive the manure.

Figure A5.3 Manure sample bag.

A5.2.4 Infrastructure provider

The manure infrastructure provider, in this case the ICT department in Vlastuin (**C**), is providing the platform for data transferring and registration (**VP**). Vlastuin has a server stack which acts as a communication centre for manure transportation (**C**). The AGR unit (Figure A5.4) sends information to the servers with GPS coordinates and scanned sample bag barcodes together with other information (**VC, C, R**). The servers (network partners (digital) internal to Vlast uin) immediately filter out only mandatory information and send this data (**value chain** functions at an internal BM in Vlastuin) to the authorities (**CU**). Authorities (**N**) send back a notification to the servers informing if the transaction was successful (external network partners' value proposition and value chain functions in BM) where it is forwarded to the AGR unit allowing further processes for manure transportation (**VP**). In the case where the transaction is not confirmed (which is very infrequent) the problem is addressed manually by calling the authorities and further addressing the problem.

The manure administrator is also connected to the server, which allows access to the laboratory results even though the laboratory (external network partner in the BM) is not connected to the servers directly. All this data can be accessed through the AGR website where the manure transporter provides additional functionalities such as Track-n-Trace (transport movement insights) and consumer specific accounting data. The AGR unit is sold with attached service contact including mobile data connection necessary for communication with the data server together with firmware updates of the unit, and software updates for the AGR website. In addition to the AGR unit, Vlastuin also provides D-Tec sampling units which takes the manure samples and packages them to the plastic bags as seen in Figure A5.3. This unit also comes with a servicing contract together with consumables and spare parts.

Figure A5.4 AGR unit.

A5.2.5 Manure administrator

The manure administrator (**network partner**) provides administrative services (network partners' **value proposition** and **value chain** function) to meet the requirements of the fertilizer law. One of the examples could be the application of manure accounting ID from the ministry (**value proposition** to user demands). The manure administrator also feeds in data from laboratory results of the manure samples. The manure administrator acts as a middle man between authorities and manure transporter, therefore; only the final data is uploaded to the authorities.

A5.2.6 Laboratory

The laboratory (**network partner**) receives the manure samples for assessment of its value. It identifies the manure producer or receiver by the barcode, and returns its findings to authorities and the manure administrator.

A5.2.7 Authorities

In this particular case, the authority is the Ministry of Agriculture and Nature Management and Fisheries (**user**) in the Netherlands. They receive the manure transporting data combined with the laboratory results (combined **value proposition**).

A5.2.8 Regulator

This is the AID (Dutch for General Inspection Service) (**user**) in the Netherlands. It ensures that all requirements are met by all the participating parties in the manure transporting process. This includes checking farmers, the manure transporter infrastructure provider, manure administrator and even the authorities themselves. If any of the requirements are violated, the violating business (or private party) is given a fine (**VP** by user).

Figure A5.5 illustrates how, on a theoretical perspective, at least two "as-is" BMs can be seen in this particular manure transportation business case. Vlastuin not only has two simultaneously operational business cases, but looking into manure transportation with just some simple business modelling details shows that the same business case – the manure transportation business case – has at least two "as-is" BMs. An overall graphical overview can be found in the illustrations in Figures A5.6, A5.7 and A5.8 of the manure loading, transportation and unloading business case.

In order to more easily understand the flow charts, the transportation processes have been split into loading, transportation and unloading.

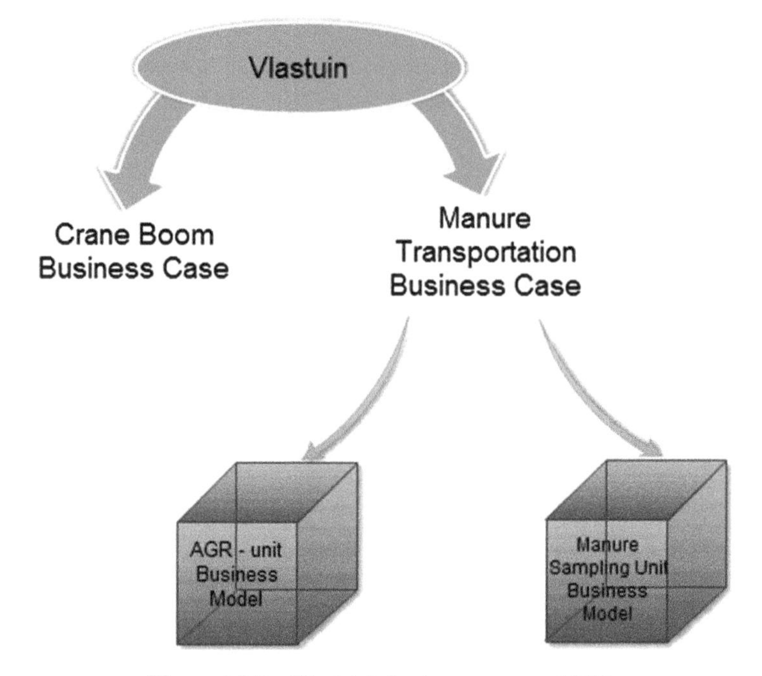

Figure A5.5 Vlastuin's business cases and BMs.

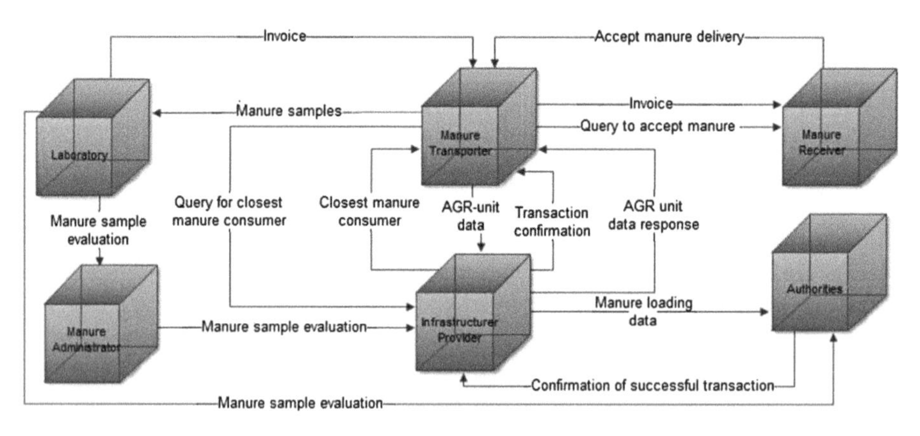

Figure A5.6 Vlastuin's business cases and BMs loading manure.

As can be seen in this very fragmented and small part of the Vlastuin's business, there are many "as-is" BMs in operation. It can also be seen that many business partners – network partners in the overview are shown each with their "as-is" BMs.

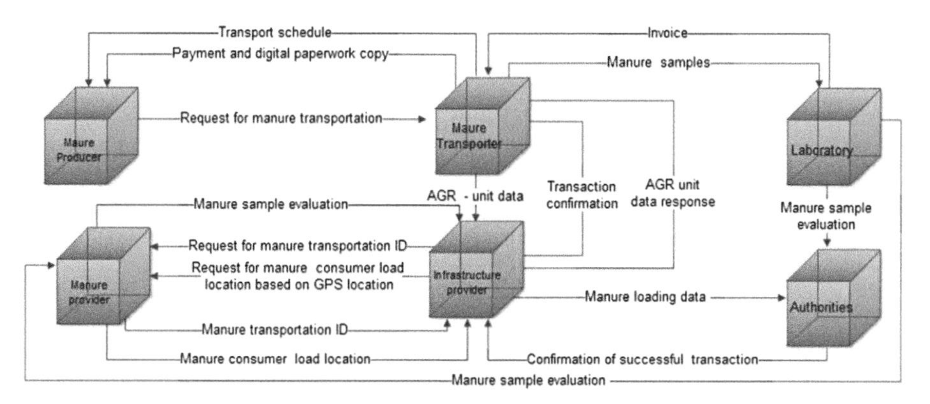

Figure A5.7 Vlastuin's business cases and BMs for manure transportation.

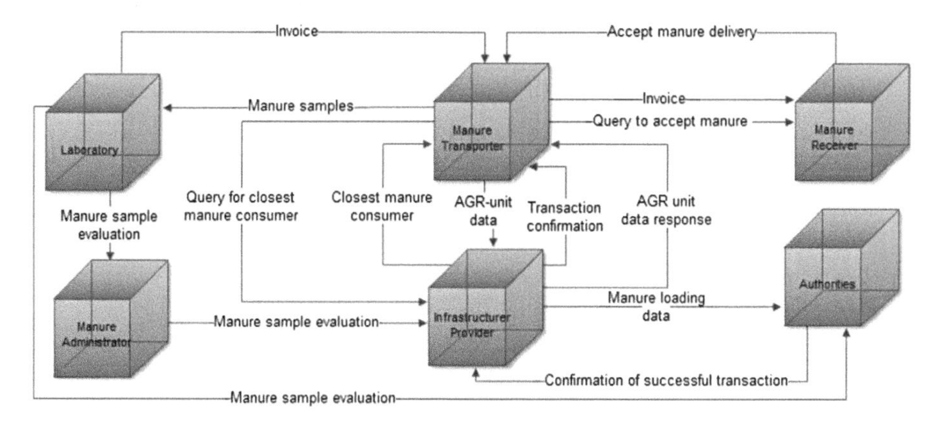

Figure A5.8 Vlastuin's business cases and BMs for unloading manure.

Appendix 6: HSJD Use Case

The hospital Sant Joan De Dieu (HSJD) belongs to the Hospital Order of Saint John of God and is a private, non-profit hospital. The order is represented in more than 50 countries and has almost 300 health care centres worldwide. HSJD is located in Barcelona, Spain, and is a children and maternity care centre. HSJD is a university hospital connected to the University of Barcelona and is also associated with the Hospital Clinic of Barcelona, which helps the hospital to provide high-level technological and patient care. HSJD is 95 per cent financed by the Catalonian public system and the remaining 5 per cent comes from private investments. The primary goal of HSJD is to encourage and educate people to follow a healthy lifestyle with good nutrition, proper sleep, hygiene and exercise.

The Risk Pregnancy (RPU) Business Case

HSJD handles and treats about 4,000 pregnancy cases per year. 10 per cent are cases where the women are at high risk of losing their babies. To postpone the birth, the doctor stops these complications and exposes the woman to a daily maternal-foetal monitoring control.

- It is real-time monitoring, concentrated in two parameters:
    - Uterine contraction
    - Foetal heart rate
- It allows the physician to view in real-time the measurement variables of the pregnant lady and her child and to take the necessary measures.
- The realization of this control involves the travelling of pregnant women to the hospital, with different frequencies of controls (some have to come every second day, others less frequently)
- It is a contradictory path: since they are high-risk patients, our physicians advise them to not move and stay calm at home. However, the control demands the pregnant women to come to the hospital every day or every two days.

Source: JJ, HSJD

In the "as-is" BM and in a number of other cases, this control involves patient's admission to the hospital. Today, it is possible to detect and measure heart rate and other key measurements from the child inside the mother. Those machines and equipment that can measure the child work very well today and nurses can do all the work on preparing and measuring the data from the child.

Today, the "as-is" BM works as the mother leaves her home for a 30-minute visit at the HSJD hospital, where a nurse generates the measurements of the child by putting the equipment on the mother's "tummy" as seen in Figure A6.1.

HSJD's doctors responsible for the pregnancy "as-is" BM find it a bit peculiar that they tell the mothers:-

"Don't do anything – do not move while at home – stay at home"

Source: HSJD Doctor responsible for RPU BM

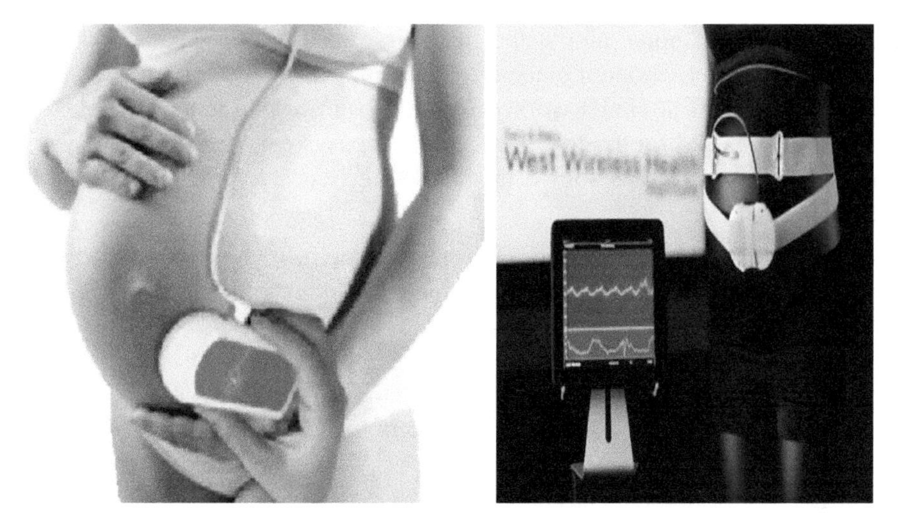

Figure A6.1 RPU “as-is” and “to-be” BM.

Figure A6.2 Diabetes patient monitoring herself at home and video and teleconference with doctor/nurse.

and then they, at the same time, ask them to transport themselves to the hospital to have the measurements done. Sometimes, the mothers have to come every second day and this is very inconvenient and not a healthy way to act, especially for those in risk of losing their child.

The doctors would therefore like to give the mothers another and better solution – something to use at home. They would like to give them some possibility to stay at home and at the same time measure the child. Today it is already possible to monitor diabetes patients in their home (Figure A6.2).

Doctors and staff at HSJD worked for two years to find technical solutions and a "to-be" BM for the challenge and BM ecosystem of risk pregnancy. The result of this work has shown the following issues seen from HSJD's perspective:

1. Cost challenge – the technique is not cheap enough. Technology has to be affordable to implement. The cost of the technology could reach $3,000 with camera, screen and so on per mother.
2. Price challenge – HSJD will not and cannot charge the mother.
3. Provider and cost challenges – West wireless institute, California US has already developed "a baby sensor" which costs US$25,000. They are interested; how interested is not known yet.
4. University of Barcelona has also developed a device but this is not tested in real environment.
5. The solution has been presented to the medical house with Philips Monitor Careview equipment; however, Philips does not want to take the risk of tele-measuring pregnancy yet.
6. Physician challenge – it is well known that the measurement can come out with false negative and false positive measurements. Doctors/physicians relying on the new device might then risk falling into some wrong conclusions.
7. HSJD is thinking about how it can involve other physicians outside – near the mother so the HSJD doctors and experts do not need to be directly involved and HSJD's "market area" can be increased.

When we were initially presented with the "to-be" risk pregnancy BM use case, we were not aware of the multitude of the "to-be" BM and BMI potential for HSJD. This was carefully studied before making the final choice and decision for one or two "to-be" BMs. Figure A6.3 illustrates the map of "as-is" BMs and the proposed "to-be" BMs registered in HSJD.

The RPU "to-be" BM is a new BMI initiative from HSJD's management which involves increasing HSJD's activities to also doing RPU with support of high-technology equipment. Therefore, this initiative involves a whole new platform of **value propositions** from HSJD, new **customers** and **users**, new **value chain functions**, new competences, new network partners, new relations and maybe new value formulae. This could be classified as, to some degree, radical innovation on many of the BM's dimensions and components. It could also address and increase the BM ecosystem for risk pregnancy as the "to-be" BM could address markets in Iraq and Morocco.

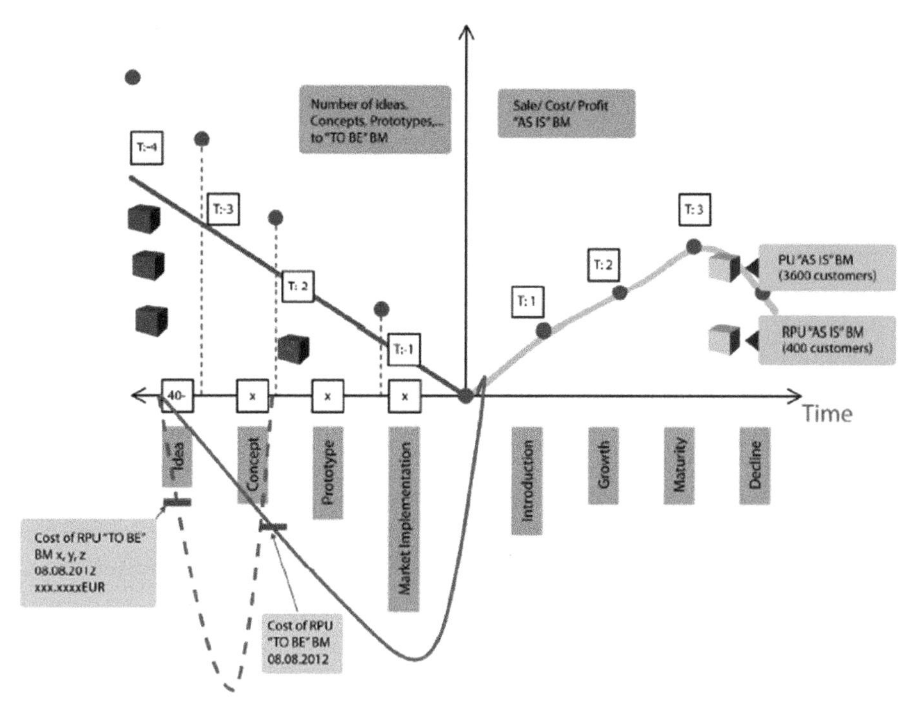

Figure A6.3 A sketch model of the BMI and BM projects in focus in RPU use case analysis related to BM and BMI lifecycle.

The RPU centre is in the "to-be" BM and in the first phase it is addressing a well-known user and customer group in Spain, but in future it would consider also addressing new **user and customer** groups external to the hospital, which would to some extent be radical related to previous target groups. We classify this change in the first phase as incremental related to most BM dimensions; however, HSJD must be aware that the customers' environment would then be outside HSJD's control and the BM would then be operating outside the HSJD physical business environment together with new network partners (tele-operators, equipment operators) which can be risky.

The **value chain set-up and functions** that have to be carried out in the RPU "to-be" BM are now related to some functions; however, HSJD has great experience in the internal and core functions of handling the functions of RPU women. The functions outside the HSJD hospital are new to HSJD and some of these are also outsourced to network partners as can be seen below.

HSJD has until now controlled most of the value chain functions around the handling of users, customers and the network in the RPU BM.

A well-developed handling programme has been tested and is operating. Now, the "to-be" BM involves other network partners. So this is all new to the HSJD pregnancy department – to some extent, a radical BMI. HSJD solved this via outsourcing some of the functions to professional network partners – for example telecomm companies, equipment providers.

- New **competences** also have to be developed for technology, HR, organizational systems and maybe also the culture. This can also mean radical innovation.
- **Network partners** were new – relations are not known, especially regarding the external network partners. However, all the relations internal in the BM are known but have to be built up from scratch. Therefore, we also classify the change on the network building block as kind of radical.
- The RPU "to-be" BM value formula is not known yet but it seems as if it may be different to other BMs in HSJD as its point of entry is related to different success criteria and different **value formulae** than profit and other BMs in HSJD.

With these characteristics we would classify the RPU "to-be" BM as seen in Table A6.1.

Table A6.1 Classification of incremental and radical BM innovation related to the seven dimensions for the RPU "to-be" BM

BM dimensions	Incremental BM innovation "Do what we do but better"	Radical BM innovation "Do something different"
1. Value proposition	Offering 'more of the same'	Offering something different (at least to the business)
2. Target users and customer	Existing market	New market
3. Value chain architecture (internal)	Exploitation (e.g. internal, lean, continuous improvements)	Exploration (e.g. open, flexible, diversified)
4. Competences	Familiar competences (e.g. improvement of existing technology, HR, organizational system, culture)	Disruptively new, unfamiliar, competences (e.g. new emerging technology, new HR skills, organizational systems, culture)
5. Network partners	Familiar (fixed) network	New (dynamic) networks (e.g. alliance, joint-venture, community)
6. Relations	Continuous improvements of existing relations (e.g. channels)	New relations, relationships (e.g. channels physical, digital, virtual, personal)
7. Value formulae	Existing processes to generate revenues and values followed-by/or incremental processes of retrenchments and cost cutting	New processes to generate revenues follow[illegible]es of retren[illegible]

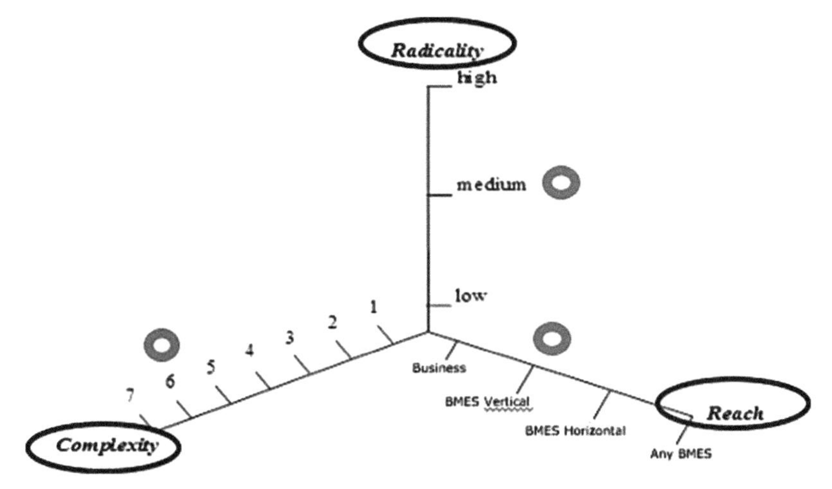

Figure A6.4 A three-dimensional business model innovation scale – risk, complexity and reach of the RPU "to-be" BM.

Seen in another diagram, the RPU's "to-be" BM could be characterized, to some extent, as a risk project as it is changing some building blocks related to the "as-is" RPU BM in HSJD seen above in Figure A6.4.

This is very much dependent on which of the several RPU "to-be" BMs HSJD would choose to implement.

In Figure 4.26, we propose the space in which the RPU "to-be" BM can be positioned in terms of its *degree* of innovativeness by means of its radicality, reach and complexity.

As can be seen, the RPU "to-be" BM is radical on innovation of BM dimensions and it is also complex as it is changing six out of seven BM dimensions. Finally, it can also be classified as far on reach as it is addressing a BM new to the business, business model vertical.

Appendix 7: BM Component List

Table A7.1 BM component list

BM dimension concept	Group of BM components	BM components
Value proposition	Product, service, process of product and service	Values, attitudes, attributes, tangible and intangible values
User and customer	A person, a family, A business	Roles
Value chain functions (internal)	Primary functions Support functions	Functions and/or activities necessary to run the BM – inbound logistics, operations, outbound logistics, marketing and sales, service – procurement, human resources management, administration and financial structure Business model innovation
Competences	Technologies HR Organizational system Culture	Product and service technologies Production technologies Process technologies Employees and people Organizational – roles Culture
Network	Physical network Digital network Virtual network	Network participants
Relations	Tangible relations Intangible relations	Relations, links
Value formula	Formulae	Formula of price and cost expressed in monetary and/or non-monetary terms

Appendix 8: List of Businesses Tested with the BMES Framework

Primary Cases in Chapter 6

AH Industries
Censec
Danish Windmill Cluster
Dong
EON
EV Metalværk
Greenlab Skive
HMN

MCH
Siemens
Vestas
Vlastuin
Watenfall

Secondary Cases in Chapter 6

Blue Ocean case research 2008
ICI Case research 2013
NEFFICS 2012
NewGibm case research 2006
SET cases 2014 and 2016
WIB 2012

Appendix 9: HSJD, Part 2

Hospital Sant Joan de Déu

Sant Joan de Déu
HOSPITAL MATERNOINFANTIL - UNIVERSITAT DE BARCELONA

A general introduction to HSJD can be found in Appendix 6, plus the RPU business case which is the basis of the findings below.

Figure A9.1 illustrates the development timeline of the hospital, and Table A9.1 gives its main facts. In 2004 HSJD experienced a need for a homogeneous profile and strategy plan caused by the growth of the hospital and the increase in activity; this resulted in the Paidhos (child in Greek) programme, which is shown on the right side of Figure A9.1, and the following values: professionalism, accessibility, innovation, teaching, hospitality, openness, solidarity and sustainability.

In 2008–2009 new needs were identified. As hospitals are typically confined to their facilities, it would open new possibilities if the patients could get in contact with the hospital by using technology like mobile phones or the internet. The Liquid Hospital (H2O) was therefore established with a goal to be an open and transparent hospital with no boundaries of facilities; it includes e-health, a social network, web portals and e-learning. Figure A9.2 illustrates the strategy of the Liquid Hospital where it is emphasized that the strategy has three action lines: the hospital, home and mobility. HSJD therefore also seeks to provide "liquid" innovative solutions for their internal and external

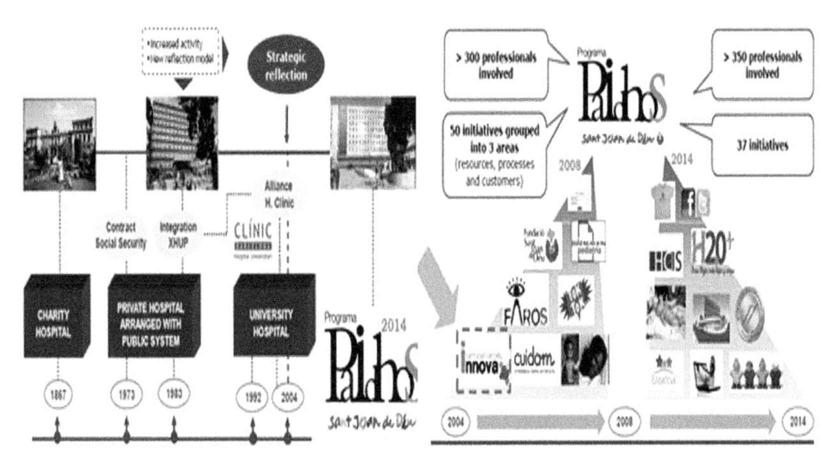

Figure A9.1 HSDJ timeline.

Table A9.1 HSDJ main facts 2010

Activity:	
Inpatient admissions	26,951
Outpatient	218,161
Emergencies visits	112,814
Surgical procedures	14,193
Childbirths	3,998
Resources:	
Staff	1,527
Beds	362
Outpatient rooms	102
Operating rooms	13
Economic data:	
Revenues M €	144.6
Investments M €	5.4

stakeholders. For example, a five-minute online training programme for their nurses educating them to use social network portals like Facebook, Twitter etc. to inform and interact with stakeholders is provided.

A9.1 Value Mapping Related to Business Value (Neffics 2012, D 3.2)

MBIT researchers commenced the work with HSJD simultaneously on two BM cases in HSDJ – the Darwin and the HVM cases – by trying to get data from HSJD on the RPU case (see Appendix 6) about the less studied segments of the value landscape (e.g. the merged world, and associated business values, value levels and value measures).

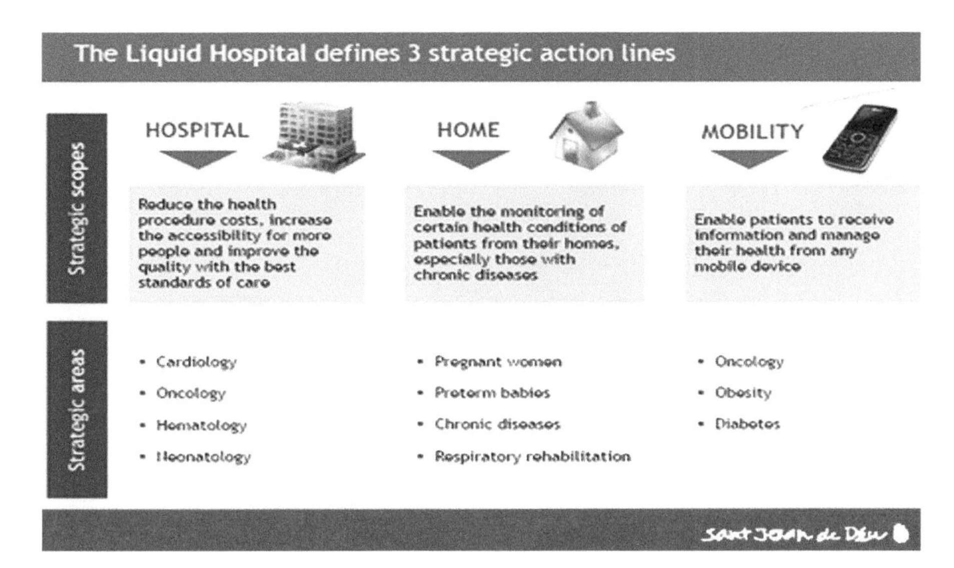

Figure A9.2 BMI strategy of the Liquid Hospital – HSJD.

The following data and value/role mapping was made available to MBIT researchers by HSJD.

From drawing value mapping such as in Figure A9.3 we were able through interview and workshops to identify more values and relations related to the stakeholders involved in the "as-is" BM. As can be seen a detailed value delivering and receiving is, however, not present, as the model shows a very overall value transfer between stakeholders in the "as-is" BM.

A more detailed value mapping and definition of value deliverables in the "to-be" BM was therefore made and it was found necessary to fully develop this. After a new information and data collection session MBIT researchers tried to set up a list of values and relate these to what values come in and go out from the stakeholders.

This was firstly done for the "to-be" BM from the viewpoint of women with risk pregnancies – "the customer leadership viewpoint" (Lindgren 2012) – and then from the the point of view of one of the other stakeholders – "the network leadership" point of view. will be shown in Part 2 of the book.

Actually many "to-be" risk pregnancy BM scenarios showed potential – however we did not have the resources to go through all "to-be" BM potentials and will therefore not show them in this appendix. A digital BMI service scenario tool would have been very helpful here to make the BMI scenario process faster. This we are now developing and will show and comment on in Part 2 of the book.

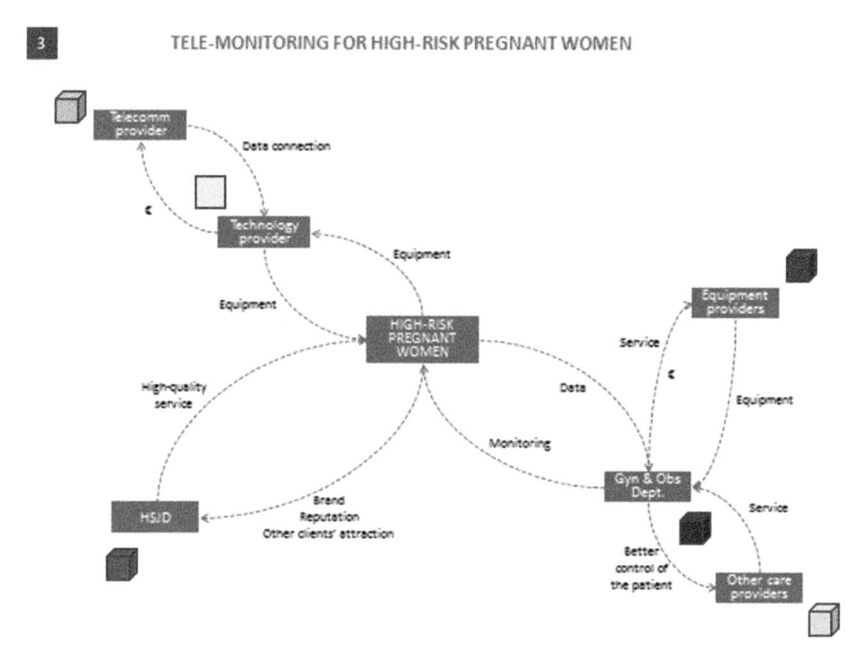

Figure A9.3 Initial value mapping sketch of Quadrant 1 – relations axiom based on data provided by HSJD's BMI department side 25 June 2012.

The woman with a high-risk pregnancy in the "to-be" BM could receive values from several identified partners – listed in the next section, who could play very different roles in the value network. However MBIT found that there could be more partners in the value mapping that could be of interest than those we original had received information from and had got information about at the interview with HSJD. Especially inside the HSJD we found that there were more stakeholders that could receive and send values to the "to-be" BMs.

A listing based on the values received by HSJD in the interview placing the high-risk pregnancy woman at the centre of the relations axiom is listed beneath. MBIT were not able at the interview to map where all the values were sent internally and externally by the woman, for example family, or to which roles – BM dimensions – they were sent.

A9.1.1 Brainstorming – first listing after downloading and seeing phase

Out – value from high-risk pregnancy woman to:

The Risk Pregnancy women send out values to several partners in the "as-is" BM network and to several BMrNs in the BM network. Below we have set up and listed those values we were able to get information about at the interview on 8 August 2012. However we believe a more detailed analysis would be able to give a much bigger value list and mapping.

A9.1.1.1 The HSJD

In – values to HSJD from Risk Pregnancy women

1. **Customer and user** – the patient is a customer and user of the HSJD Pregnancy Centre.
   Roles – customer and user
2. **Getting/recruiting branding to other users/customers** for the HSJD Pregnancy Centre – capturing customers and marketing HSJD to other pregnant or potentially pregnant women.
   Roles – "selling" role, capturing users and customer role, "branding provider"
3. **HSJD gets paid by the government for each consultation**
   Roles – "indirect payment via the government"

Out – value from HSJD to the patient

On the basis of the study the following "to-be" BMs could be identified

1. **"To-be" BM 1: Pregnancy service** – the woman gets all pregnancy health care services from HSJD – right from first consultation to birth to post-natal care. This means that she demands a process of value propositions – with different value propositions (products, services and processes) "tailor made" to the individual woman aligned with where she is in the value proposition process.
   Roles – health care provider – before, during and after pregnancy
2. **"To-be" BM 2: Health care security** – HSJD provides the best available health care for the pregnant woman – security for the RPU patient, her child and her family.
   Roles – health care security provider
3. **"To-be" BM 3: Pregnancy data service** – the woman gets all the pregnancy health care data services demanded by government and others provided by HSJD – right from first consultation to birth to post-natal care. This is a process of value propositions – data registration and services "tailor made" to the individual woman where she is in the process.
   Roles – health care provider – before, during and after pregnancy
4. **"To-be" BM 4: Pregnancy funding service** – the woman gets all the pregnancy health care funding data services demanded by government and others to fund and support medicine provided by HSJD – right from first consultation to birth to post-natal care. This is a process of

value propositions – data registration and services "tailor made" to the individual woman and where she is in the value proposition process.
Roles – health care funding service provider – before, during and after pregnancy

A9.1.2 The medical technology provider's viewpoint – network business model innovation leadership

In – value to Risk Pregnancy woman from medical technology provider

1. Provide safe and secure equipment that the Risk Pregnancy woman can use
   Roles – safe and secure equipment provider/supplier
2. Provide maintenance and service of equipment to the Risk Pregnancy woman
   Roles – maintenance and service provider/supplier
3. Provide teaching/training and teaching/training material for the Risk Pregnancy woman – train the trainers and the service personnel at the Darwin Centre (http://simpeds.org/international_coe/about-barcelona/)
   Roles – training role if this has to be done by medical staff

Out – value to medical technology provider from Risk Pregnancy woman

1. Provide a user and/or customer base for the medical technology provider businesses
   Role – user, customer, showcase user and/or customer, reference role.
2. Medical businesses can have users and customers as showcases, where other potential users and/or customers can see the use of the equipment and thereby develop a preference for the equipment
   Roles – use and/or customer showcase role, reference provider, brand provider

The medical technology provider to HSJD

In – value to HSJD from medical technology provider

1. Provide safe and secure equipment that the Risk Pregnancy woman can use – tangible value delivered to HSJD
   Roles – safe and secure equipment provider/supplier
2. Provide maintenance and service of equipment to the Risk Pregnancy patient – always maintained and serviced equipment – tangible value to HSJD
   Roles – maintenance and service provider/supplier though HSJD
3. Provide teaching and teaching material for the HSJD staff and maybe also for the Risk Pregnancy women – intangible/tangible value as

high quality and pedagogical teaching/training and teaching/training material – train the trainers (HSJD personnel or others) so that this service can be provided by the personnel at the HSJD hospital or other centres to pregnant women
Roles – direct and indirect teaching and training provider role

Out – value to medical technology provider from HSJD

1. Provide a user and/or customer base for the medical technology provider businesses
   Roles – user and customer provider, showcase user and/or customer, reference role
2. Medical companies can have a users' and customers' showroom, where other potential users and/or customers (B2B hospitals) can see the use of the equipment and thereby develop a preference for it.
   Roles – use and/or customer showcase role, reference provider, brand provider

A9.1.3 The telecomm provider's viewpoint

In – value to HSJD from telecomm provider

1. Provide safe and secure telecomm equipment, line, bandwidth that the Risk Pregnancy woman can use – tangible value delivered to HSJD
   Roles – safe and secure equipment provider/supplier
2. Provide maintenance and service of telecomm connection to the Risk Pregnancy equipment – always maintained and serviced equipment – tangible value to HSJD
   Roles – maintenance and service provider/supplier to HSJD
3. Provide teaching and teaching material for the HSJD staff and maybe also for the Risk Pregnancy women if necessary on the telecomm service – intangible/tangible value as high quality and pedagogical teaching/training and teaching/training material – train the trainers (HSJD personnel or others) so that this telecomm service can be provided by the personnel at the HSJD hospital or other centres to pregnant women
   Roles – Direct and indirect teaching and training provider role

Out – value to telecomm provider from HSJD

1. Provide a user and/or customer base for the telecomm provider businesses
   Roles – user and customer provider, showcase user and/or customer, reference role
2. Telecomm provider can have a users' and customers' showroom, where other potential users and/or customers (B2B hospitals) can see the use of the telecomm service and thereby develop a preference for it

Roles – user and/or customer showcase role, reference provider role, brand provider role

A9.1.4 The Gyn. and Obs. Department

In – value to Gyn. and Obs. Department from Risk Pregnancy women

1. Provide centre with participants – users/and customers. The Risk Pregnancy women provide these centres with users/and customers.
   Roles – customer and user of the Gyn. and Obs. Department

Out – value from Gyn. and Obs. Department to Risk Pregnancy women

1. Branding and reference provider for Gyn. and Obs. Department
   Roles – branding provider, reference provider
2. Course and training provider to Risk Pregnancy women
   Roles – course and training provider

The previous value mapping is not complete but has been described around stakeholders near to one of the Risk Pregnancy "to-be" BMs. However, in another BM case the Darwin Case customers also receive and send values to the customer, suppliers and others from outside HSJD hospital.

On the basis of the detailed value description it was possible to draw the following revised value mapping of the "to-be" Risk Pregnancy BM. In this revised value mapping it is possible to see slightly more detailed values – however still not the complete value mapping of tangible and intangible values, nor all roles around the "to-be" Risk Pregnancy BM. In Figure A9.4 we have tried to model the different BMs that are potentially involved in the Risk Pregnancy Unit (RPU) BM. As it is a Network based (NB) BM these BMs have to be "merged" together to the new RPU "to-be" BM.

To form the new RPU "to-be" BM a forming process has to take place as we illustrate in Figure A9.5.

To do the value mapping, a deeper analysis was necessary to get more information from HSJD, which was not available on 8 August 2012. However it was stated by "JJ" at the hospital that HSJD and the innovation department at HSJD do three value metrics as mentioned above. This also includes the Risk Pregnancy "to-be" BM.

In the interview and workshop we also saw some attempts to register costs of BMI Risk Pregnancy project – especially operating cost, costs that HSJD will have to manage and service, for example the Risk Pregnancy "to-be" BM. However, we did not see any calculations of proposed "to-be" business models.

The business value in the "as-is" BM for Risk Pregnancy and the "to-be" BM can indeed, as with the other BM use cases, be variously described,

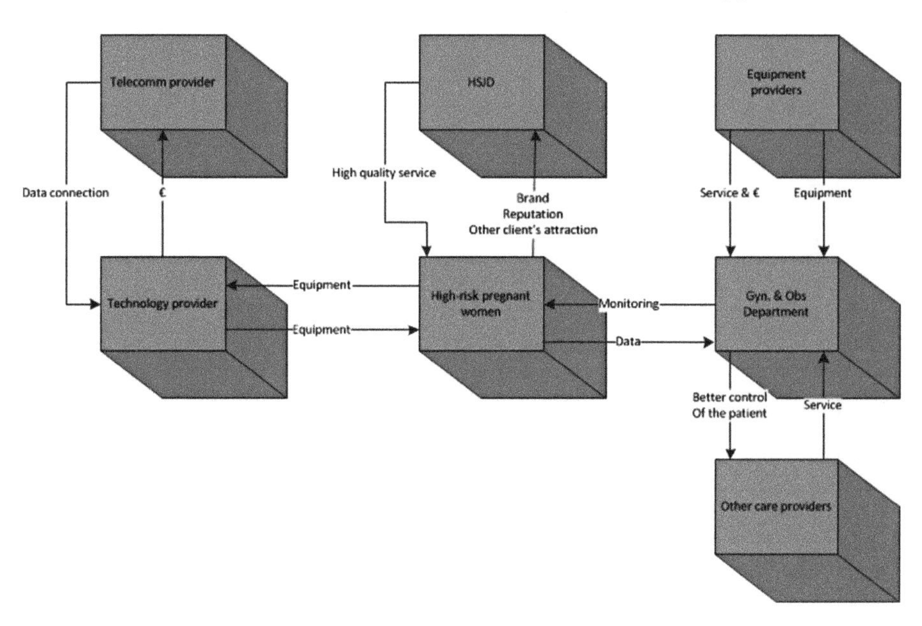

Figure A9.4 "As-is" and "to-be" BMs involved in forming the new RPU "to-be" BM.

interpreted, measured and used on different levels. However the detailed information was not available to the MBIT research group on 8 August 2012 and it would take a great deal of time and effort for HSJD to get it.

In this context even if the value agents and the value activity involved are the same, there could be very different outcomes of the value activity depending on the level at which it was analysed, and again it would take time and effort for HSJD to make this information available.

However, we provided and prepared a first table (Table A9.2) of the different value levels for the Darwin Case to be filled out by HSJD.

It can be stated that this work has to be done for every possible Risk Pregnancy "to-be" BM to get the full value potential scenario picture. As there are many different potential "to-be" BMs in the Risk Pregnancy use case as the MBIT researchers saw it this would take some time to develop. In this case again a BMI service tool for helping HSJD in this work would be highly valuable.

Table A9.2 Value levels related to the Darwin "as-is" and "to-be" BM

LEVEL	DESCRIPTION	DESCRIPTION "AS-IS BM"	DESCRIPTION "TO-BE BM"
L1	**Value at the level of the individual business (typically a business)** This is typically (defined) within the business model of a firm and is expressed through the value proposition of a particular business (value creation).	**Value proposition today**	**Value Proposition "TO BE" N1, N2 . . . NN**
L2	**Value at the level of the business model ecosystem (BMES)** This is correlated to the BMES within which a business operates and is expressed through the value of an offering (one or more value objects) to the market for the offering encompassing both its supply and demand.	**Users** **Customers** **Suppliers** **Network partners** **Other**	
L3	**Value at the level of the whole business model ecosystem (BMES)** This is correlated to business ecosystem as a whole where the focus is on the value of an offering relative to those of other offerings in the same or other related BMESs – either vertical or horizontal BMESs, as part of a larger "value package" that drives supply and meets demand in aggregates.	**Hotel**	
L4	**Value at the societal level** This is correlated to the broader societal value as determined by culture and social norms and other drivers.	**Spain** **Africa**	

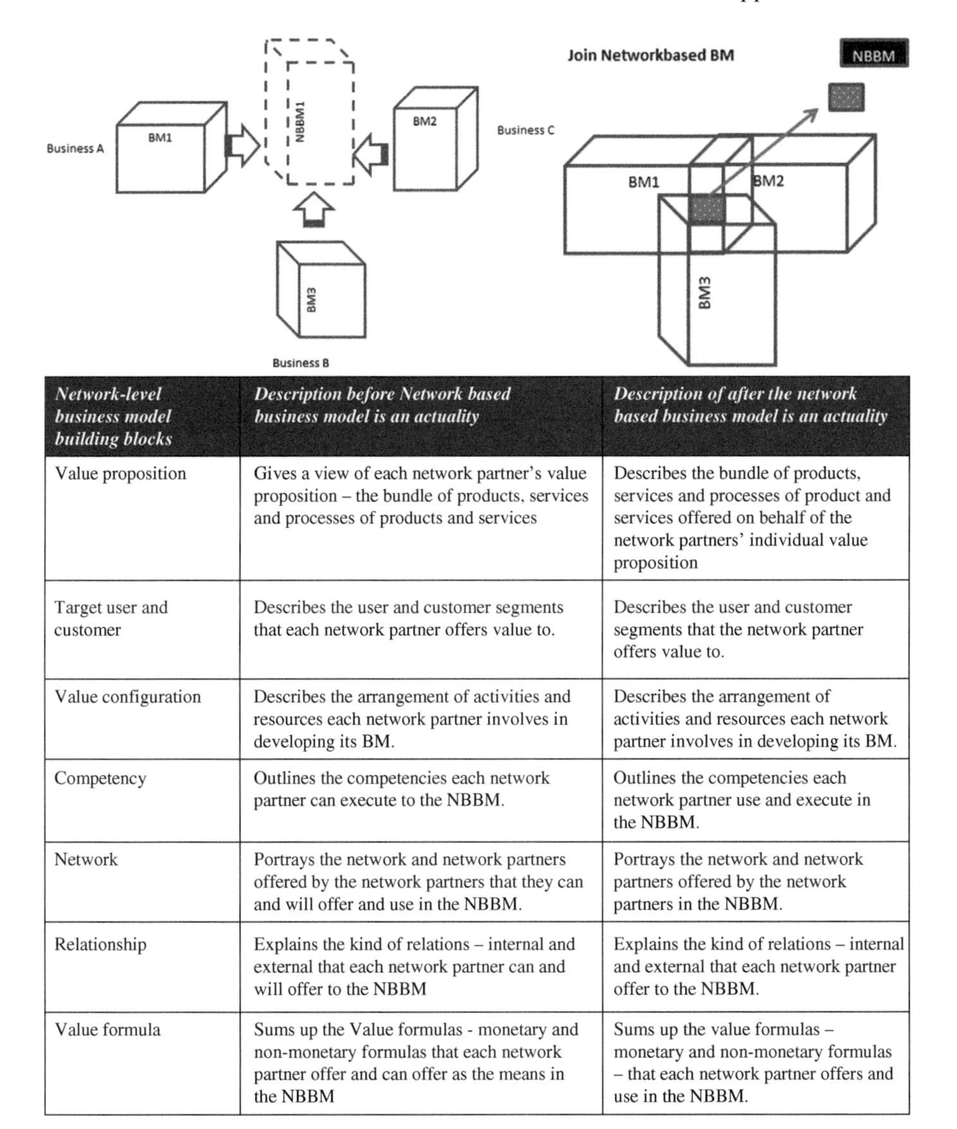

Network-level business model building blocks	*Description before Network based business model is an actuality*	*Description of after the network based business model is an actuality*
Value proposition	Gives a view of each network partner's value proposition – the bundle of products, services and processes of products and services	Describes the bundle of products, services and processes of product and services offered on behalf of the network partners' individual value proposition
Target user and customer	Describes the user and customer segments that each network partner offers value to.	Describes the user and customer segments that the network partner offers value to.
Value configuration	Describes the arrangement of activities and resources each network partner involves in developing its BM.	Describes the arrangement of activities and resources each network partner involves in developing its BM.
Competency	Outlines the competencies each network partner can execute to the NBBM.	Outlines the competencies each network partner use and execute in the NBBM.
Network	Portrays the network and network partners offered by the network partners that they can and will offer and use in the NBBM.	Portrays the network and network partners offered by the network partners in the NBBM.
Relationship	Explains the kind of relations – internal and external that each network partner can and will offer to the NBBM	Explains the kind of relations – internal and external that each network partner offer to the NBBM.
Value formula	Sums up the Value formulas - monetary and non-monetary formulas that each network partner offer and can offer as the means in the NBBM	Sums up the value formulas – monetary and non-monetary formulas – that each network partner offers and use in the NBBM.

Figure A9.5 Model of the network based RPU "to-be" BM.

Appendix 10: Empirical BM Relations Cases: Quadrant 1 – the First Square of the Relations Axiom – "in in BM Relations"

Margit Gade Case – ***Example of Quadrant 1 of the Relations Axiom***
Margit Business is a Danish start-up business within the learning industry. Margit Business develops learning material – books, homework portals, courses – for primary school children who face difficulties in understanding existing learning materials. Margit Business translates existing material into learning materials and teaching materials that help children to learn and learn better, and supports teachers to teach these children. The Margit Business operates in Denmark, Norway and Netherlands. Margit Business develops and produces all material together with elected network partners.

The Margit case showed us a business that had many tangible and intangible BM relations but kept in a very closed network and particularly inside the business. Margit struggled in establishing her business, pushing the release of BM competences to find new BMs. Margit was, when we entered the project, mainly focusing on internal competences in her BMI but when we later finished the project was continuously developing her relations and use of relations to customers and network partners. The method released competences and hidden IC/knowledge both internal to the business model and between business models within the business and external to the business.

Margit Business was included in a business model innovation camp run by MBIT researchers – "the Junget Business Model Innovation Camp" – together with five other business from different sectors. The businesses were encouraged to find, tell and visualize their relations and business models at a social, human, intellectual capital and personal level and in their relations to competences and intellectual capital (IC). All these tangible and intangible relations were considered important to a business and its hidden business model innovation capacity – which many in traditional business model innovation and traditional business model innovation capacity measurements do not see and often forget to work with and focus on. Picture A10.1 shows the mapping process in the Junget Camp including Margit Business mapping her relations to competences and IC as can be seen in Figure A10.1.

Methods of releasing BM value, competences, intangible IC and hidden knowledge in the Junget Camp – and the other cases – was based on a carefully planned process including different tools, processes and expression methods (storytelling, discussion, writing, 3M noting, BM relations axiom mapping,

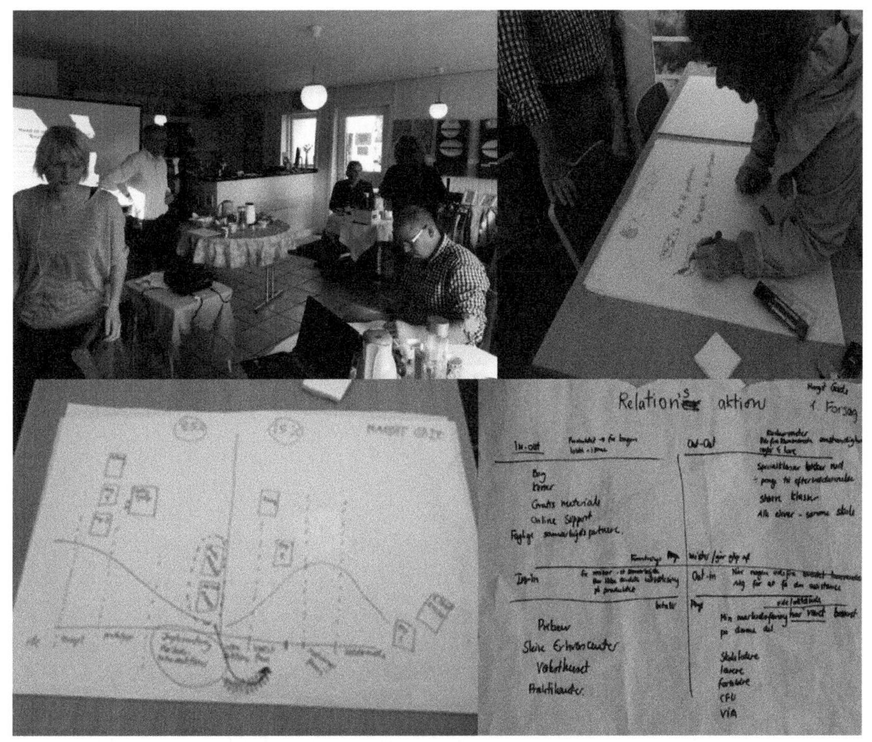

Picture A10.1 Pictures from the Junget Business Model Innovation Camp releasing competences, IC and hidden knowledge in business model relations.

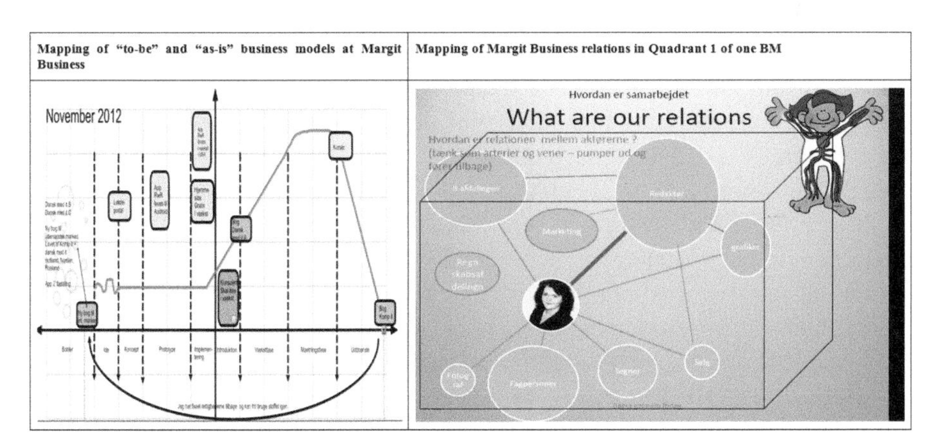

Figure A10.1 Margit Business BM use case.

walk and talk, two by two and group discussion and processes, homework, pitching, rapid prototyping and presentation tools). The difference in methods was chosen because MBIT researchers realized that different businesses measure relations in different ways. This helped us to release very powerful different values, competences, intangible IC and hidden knowledge in the businesses both in the single sessions we had with the businesses but also over the whole process. Each of the businesses "found" and "discovered" relations to far more potential competences and IC to do BMI than they were even able to use fully in their business. This stressed again the importance of mapping and having an overview of one's business's BMs relations to IC.

Appendix 11: The EV Metalværk Case – Quadrants 1, 2 and 3

EV Metalværk is a Danish business producing fittings and valves. EV Metalværk is a total supplier of turned and milled parts with high technology production equipment. It offers its products in many different materials: brass, steel, stainless steel, aluminum, titanium, plastic etc. and produces, together with qualified subcontractors, any kind of surface treatment. EV Metalværk works as subcontractor to more than 50 different branches. EV Metalværk offers varied production methods – including CNC turning, CNC milling, long turning and different mounting assignments. It operates in the EU and the US. It holds a niche position in the high and middle end valve OEM market. EV Metalværk's own production and innovation focuses on the high to middle pressure market.

The EV Metalværk use case showed us that a business and business models had many tangible and intangible relations. Especially in Quadrants 1, 2 and 3 their relations mapping looked, as they described it, like "spaghetti", as shown in Figure A11.1. Due to lack of external networking and internationalization EV Metalværk had not much to map in Quadrant 4. They did not know much about potential IC that could contribute to their business models. They were hardly using or focusing on Quadrant 4.

As EV Metalværk was mainly focusing on internal competences, IC and relations to well-known customers and network partners in Quadrants 1, 2 and 3 the business BMI was strongly based on internal and narrow competences.

A simple map of EV Metalværk's relations to competence and IC in BMs gave valuable inputs to where they could release competence and IC and network in their BMs. However, as seen in Quadrant 3, EV Metalværk needed more information as to where in the business models outside the business EV

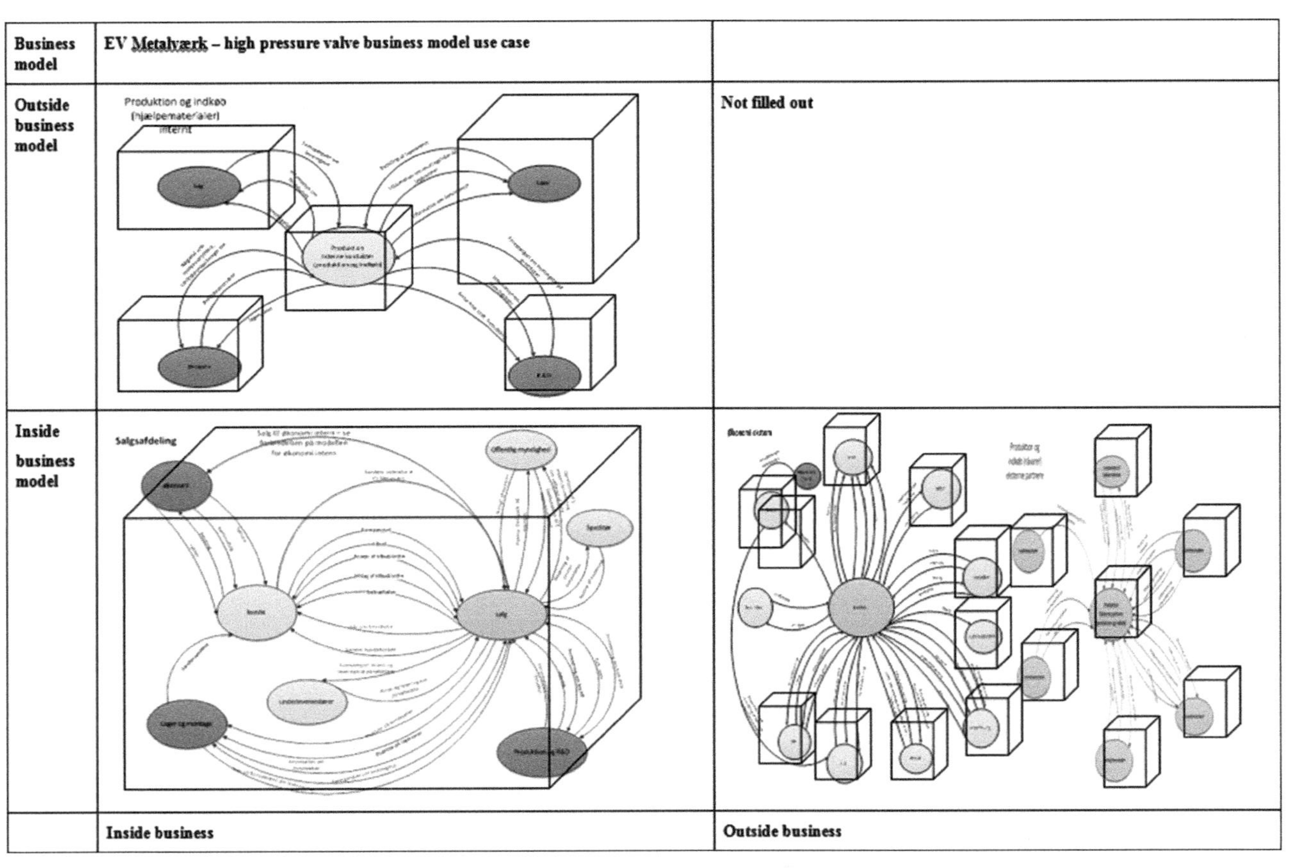

Figure A11.1 EV Metalværk business use case.

Metalværk's BM relations were linked to. This demanded a deeper insight into the BMs outside EV Metalværk's business.

Appendix 12: Quadrant 4 – The Fourth Square of the Relationship Axiom – "out out" Relations

For a further understanding and exemplification of the relations axiom, the business of a no-frills airline, e.g. Ryanair, will serve as a practical example (Table A12.1).

Ryanair has many business models, for example the flights themselves, different sales BMs, several fee BMs, and many more. Besides, it collaborates with other businesses, for example "third party providers": in Ryanair's terms, airports, and aviation fuel and lubricants suppliers. Crewlink is one of the businesses which provides cabin crew to Ryanair. Stansted Aircraft Maintenance Services is one of the businesses which provide maintenance to Ryanair's aircrafts. Aviation fuel supply businesses, e.g. Shell Aviation, supply fuels and lubricants to aircrafts.

Airport Hamburg has four main business areas: aviation, ground handling, centre management, and real estate management. For the first two business areas, some examples of BMs are given in the following. The main task of aviation is to ensure the smooth operation of the airport in cooperation with the airlines, by securing the airport logistics, and providing security services, a plant fire brigade, and other services. Ground handling is divided into four businesses: AHS Aviation Handling Services, CATS Cleaning and Aircraft Technical Services GmbH & Co. KG, GroundSTARS, and STARS, "Special transport and ramp services". Each of those has several BMs; a selection is shown in Table A12.1.

The following shows only an excerpt of the businesses and business models which are related to the operation of a no-frills airline like Ryanair.

Firstly we show in Figure A12.1 one of Ryanair's business models in Quadrant 1.

Inside the business of the no-frills airline and outside the focal BM flight, relations of the focal BM to the BM sales of food, beverages, confectionery, on-board snacks (BMSF) and to the BM of retail shopping, i.e. sales of fragrances, skincare, cosmetics, jewellery and watches (BMRS) are shown in Figure A12.2.

BMF is providing sales competence (lighter arrows) from its competence dimension and is selling the catering items for BMSF, and the retail items for BMRS. BMSF and BMRS in turn make internal payments (darker arrows) from their value formula dimension to the value formula dimension of BMF.

Table A12.1 Excerpts of BM in aviation

No-frills airline, e.g. Ryanair	Third party providers	Airport, e.g. Hamburg, Germany	Aviation Fuel e.g. Shell Aviation
Flights (BM_F)	Crewlink	Aviation	Aviation fuel (BM_{AF})
Sales of food, beverages, candies, snack on board (BM_{SF})	Cabin crew (BM_{CC})	Airport logistics	Aviation lubricants
Retail Shopping, i.e. sales of fragrances, skincare, cosmetics, jewelry, and watches (BM_{RS})		Plant fire brigade	
Sales of scratch-cards	Stansted Aircraft Maintenance Services	Security	
Fees for surplus baggage		FH Ground Handling	
Fees for equipment	Engineering services	*AHS Hamburg*	
Fees for fast track		Check-In	
Fees for Business Plus		IATA Travel agency	
Sales of Samsonite bags		*CATS*	
		Cleaning (BM_C)	
		GroundSTARS	
		Loading, unloading	
		Baggage handling	
		Technical equipment provision	
		STARS	
		Passenger transport to and from aircraft (BM_{PT})	
		Crew transport to and from aircraft	
		Pushing aircraft into desired positions on airfield & parking	

As these are contracted value exchanges, the arrows are shown with continuous lines. Intangible relations transferring non-contracted values are shown with dotted lines, for example information from BMRS to BMF, and from BMF to BMSF.

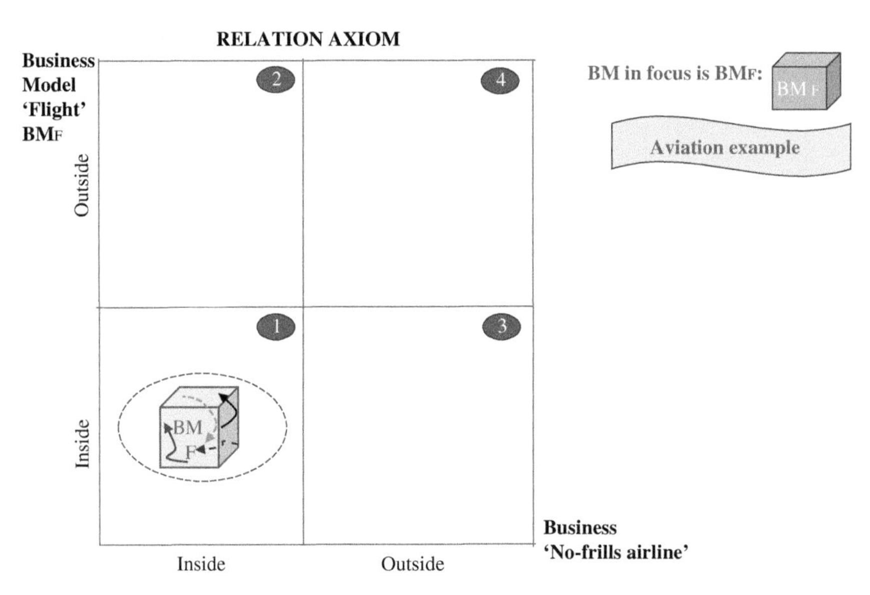

Figure A12.1 Quadrant 1 Ryanair business model.

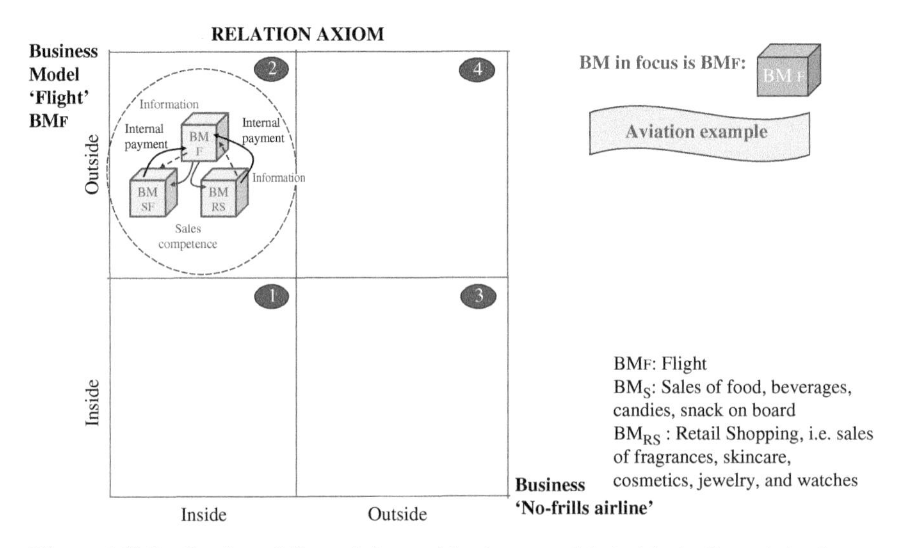

Figure A12.2 Quadrant 2 Ryanair internal business models inside the Ryanair business.

In Figure A12.3 it is possible to see business models of Ryanair that are operating outside the business.

Finally we show in Figure A12.4 some business models that Ryanair is not involved in and is not interacting with outside the business.

Summing up, we get a full relations axiom picture of an elected part of Ryanair business as seen in Figure A12.5.

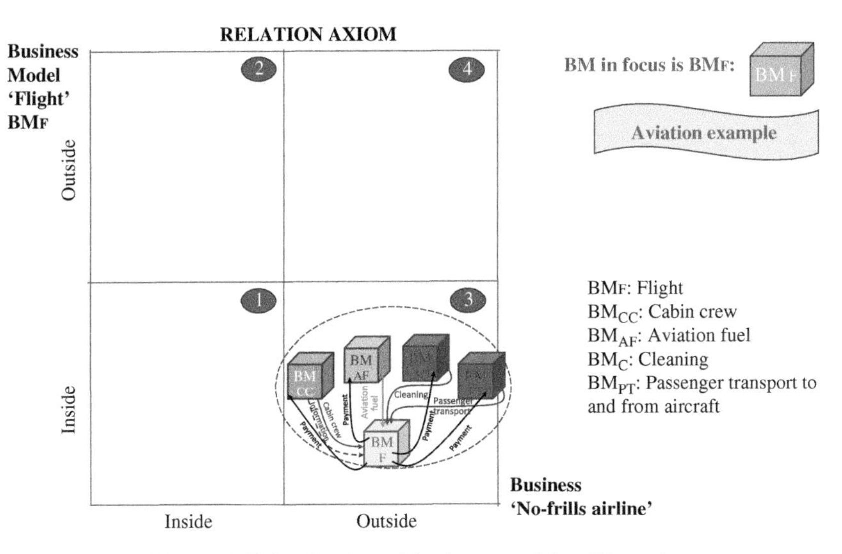

Figure A12.3 Quadrant 3 business models of Ryanair.

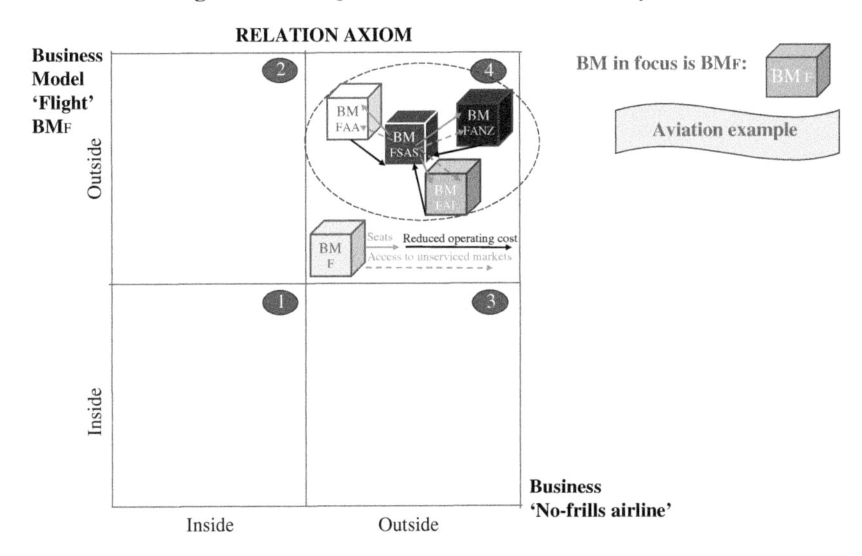

Figure A12.4 Quadrant 4 business models that Ryanair is not part of.

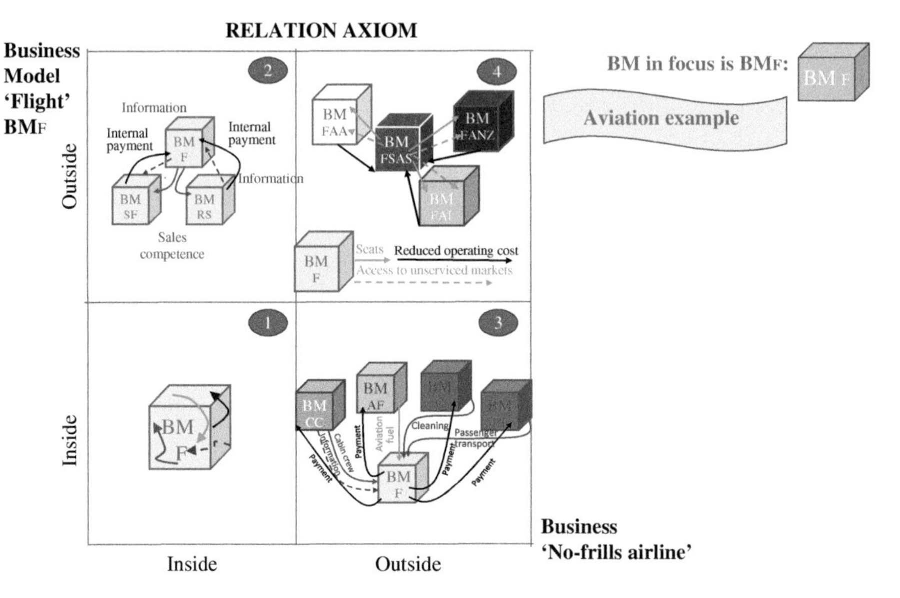

Figure A12.5 Full relations axiom picture of a part of Ryanair's business.

References

Abell, D. F. (1980). Defining the Business: the Starting Point of Strategic Planning (Upper Saddle River, NJ: Prentice-Hall, Inc.).

Abdullah, A. M. and Lindgren, P. (2008). "Innovation Leadership in Danish SMEs Management of Innovation and Technology," *ICMIT 2008. 4th IEEE International Konference. IEEE, 2008.*

Afuah, A. and Tucci, C. (2003). *Internet Business Models and Strategies*, Boston, McGraw Hill.

Albrecht, K. (1992, November). The Only Thing that Matters, *Executive Excellence*, 9, 7.

Alderson, W. (1957). Marketing Behavior and Executive Action (Homewood, IL: Irwin).

Allee, V. (2008). "Value Network Analysis and Value Conversion of Tangible and Intangible Assets," *Journal of Intellectual Capital*; 9, 1: 5–24.

Allee, V. (2010). VDML and the Value Networks View. Presentation on the OMG 2010 Conference at Santa Clara California USA.

Allee, V. (2012). Definition of Value Capacity. Available at: http://valuenetworks.com/public/item/236677.

Allee, V. and Schwabe, O. (2011). Value networks and the true networks of collaboration Open Source http://www.valuenetworksandcollaboration.com/.

Amidon, D. M. (2008). Innovation SuperHighway (Amsterdam: Elsevier Science).

Amit, R. and Zott, C. (2001). "Value Creation in e-business," *Strategic Management Journal*, 22, 6–7: 493–520.

Anderson, J. C. and Narus, J. A. (1999). Business Marketing Management: Understanding, Creating and Delivering Value (Upper Saddle River, NJ: Prentice Hall).

Anderson, P. (1982). Marketing, Strategic Planning and the Theory of the Firm. *Journal of Marketing*; 46, 2: 15–26.

Austin, R. and Devin, L. (2004). Artful Making: What Managers Need to Know About How Artists Work (Financial Times Prentice Hall).

Axelsson, B. and Easton, G. (1992). Industrial Networks: A New View of Reality (Routledge).

Baden-Fuller, C. (2015). Business models and modelling, in: Advances in Strategic Management (Emerald Insight).

Baden-Fuller, C. and Haefliger, S. (2013). "Business Models and Technological Innovation," *Long Range Planning*; 46: 419–426.

Baden-Fuller, C. and Mangematin, V. (2013). "Business Models: A Challenging Agenda," *Strategic Organization*; 11, 4: 418–427.

Blois, K. (2004). "Analyzing Exchanges through the Use of Value Equations," *The Journal of Business & Industrial Marketing*; 19, 4–5: 250–257.

Bouwman, H. (2003). Designing Metrics for Business Models Describing Mobile Services Delivered by Networked Organisations. In *16th Bled Electronic Commerce Conference eTransformation*, 1–20.

Bouwman, H., De Vos, H. and Haaker, T. (2008). Mobile Service Innovation and Business Models (Springer).

Bower, J. L. and Christensen, C. M. (1995). "Disruptive Technologies: Catching the Wave," *Harvard Business Review*; 73, 1: 43–53.

Brodie, R. J., Brookes, R. W. and Coviello, N. E. (2000). "Relationship marketing in customer markets," in *The Oxford Textbook on Marketing*, ed. Blois, K. (Oxford: Oxford University Press), 517–533.

Caffyn, S. and Grantham, A. (2003). "Fostering Continuous Improvement within new Product Development Processes," *International Journal of Technology Management*; 26: 843–878.

Casadesus-Masanell, R., and Ricart, J. E. (2009). "From Strategy to Business Model and to Tactics," Working Paper. Harvard Business School. Available at: http://www.hbs.edu/research/pdf/10-036.pdf.

Casadesus-Masanell, R. and Ricart, J. E. (2010). "From Strategy to Business Models and onto Tactics," *Long Range Planning*; 43: 195–215.

CGC (2006). CTIF Global Capsule. Available at: http://www.ctifglobalcapsule.com/organization/.

Chan, K. W. and Mauborgne, R. (2005). Blue Ocean Strategy. How to Create Uncontested Market Space and Make Competition Irrelevant (Boston, MA: Harvard Business School Press).

Chapin, F. S., Matson, P. A. and Mooney, H. A. (2002). Principles of Terrestrial Ecosystem Ecology (New York: Springer).

Chesbrough, H. (2007). Open Business Models How to Thrive in the New Innovation Landscape (Brighton, MA: Harvard Business School).

Chesbrough, H. (2010). "Business Model Innovation: Opportunities and Barriers," *Long Range Planning*, 43: 354–363.

Chesbrough, H., Vanhaverbeke, W. and West, J. (2008). Open Innovation: Researching a New Paradigm (Oxford University Press).

Child, J. and Faulkner, D. (1998). Strategies of Co-operation – Managing Alliances, Networks, and Joint Ventures (Oxford: Oxford University Press).

Child, J., Faulkner, D. and Tallmann, S. (2005). Cooperative Strategy – Managing Alliances, Networks and Joint Ventures (Oxford: Oxford University Press).

COIN (2010, 31 January). "Deliverable D6.2.1a – Integrated EI Value Proposition," M24 issue. Available at: http://www.coin-ip.eu/research/coin-results/public-documents/all-public-documents/COIN%20D6.2.1a%20 First%20Integrated%20EI%20Value%20Proposition%20v2.0-FINAL. pdf/view.

Cold Hawaii (2016). Available at: http://www.getwetsoon.de/coldhawaii/ and http://www.yourdanishlife.dk/never-been-to-hawaii-check-out-the-waves-of-cold-hawaii-instead/.

Cooper, R. (1993). Winning at New Products (Reading, MA: Addison-Wesley Publishing Company).

Cooper, R. G. (2005). Product Leadership: Pathways to Profitable Innovation, 2nd ed. (New York: Basic Books).

Coviello, N. E., Brodie, R. J., Danaher, P. J. and Johnston, W. J. (2002). "How Firms Relate to Their Markets: An Empirical Examination of Contemporary Marketing Practices," *Journal of Marketing*; 66, 3: 33–46.

Cowi (2010). International Center for Innovation, Aalborg Universitet. *Midtvejsevaluering af Internationalt Center for Innovation Evalueringsrapport Oktober 2010 COWI Dokumentnr. A012527 Version 1 Udgivelsesdato 8. Oktober 2010* Udarbejdet NIGE, BGJE Kontrolleret AR Godkendt NIGE.

Daft, R. L. (2010). Understanding the Theory and Design of Organisations (Vanderbilt: South-Western).

Danish Ministry of Climate and Energy. (2011, 24 February). "Release of Danish Energy Strategy 2050". Available at http://dfcgreenfellows.net/ Documents/EnergyStrategy2050_Summary.pdf.

Dansk Statisk. (2012, 20 March). Main Energy Statistics 2011. Danish Ministry of Climate and Energy.

Davenport, T. (1990). The New Industrial Engineering: Information Technology and Business Process Reengineering (MIT Sloan Management Review).

Davies, R. (2009). The Use of Social Network Analysis Tools in the Evaluation of Social Change Communications. An input into the Background Conceptual Paper: An Expanded M&E Framework for Social Change Communication.

Day, G. G. (2000). "Managing Market Relationships," *Journal of the Academy of Marketing Science*; 28, 1: 24–30.

Dougherty, D. (1992). "Interpretive Barriers to Successful Product Innovation in Large Firms," *Organisation Science*; 3, 2: 179–202.

Doyle, P. (2000). Value-based Marketing: Marketing Strategies for Corporate Growth and Shareholder Value (Chichester: John Wiley and Son).

Drucker, P. (1973). Management: Tasks, Responsibilities, Practices (New York: Harper Rose).

DWI (2014). "Read the Danish Wind Industry Annual Statistics 2014," Report – in Danish only.

EU (2017). Regions & Cities of Europe. Available at: http://cor.europa.eu/en/news/regions-and-cities-of-europe/Documents/3095-magazine-CoR-98.pdf.

Fielt, D. E. (2011). "Conceptualising Business Models: Definitions, Frameworks and Classifications," *Journal of Business Models*; 1, 1: 85–105.

Flarup, J., Jensen, S. S. and Lindgren, P. (2016). "Business Model Innovation Competences: What Interdisciplinary Competences Can Really Value and do Business Model Innovation," *Paper presented at Global Wireless Summit 2016, Aarhus, Denmark.*

Fogg, B. J. and Kaufmann, M. (2003). Persuasive Technologies – Using Computers to Change what We Think and Do (San Francisco, CA: Morgan Kaufmann).

Ford, D. (ed.) (2001). Understanding Business Marketing and Purchasing. 3rd ed. (London: Dryden Press).

Ford, D., Berthon, P., Brown, S., Gadde, L.-E., Håkansson, H., Naudé, P., et al. (2002). The Business Marketing Course: Managing in Complex Networks (Chichester: John Wiley & Sons).

Ford, D., Håkansson, H., Gadde, L.-E. and Snehota, I. (2003). Managing Business Relationships. 2nd ed. (Chichester: John Wiley & Sons).

Freeman, L. C. (2009). The Development of Social Network Analysis—with an Emphasis on Recent Events. Available at: http://moreno.ss.uci.edu/91.pdf.

Friedman, T. (2007, November). "The World is Flat 3.0," *Lecture at Massachusetts Institute of Technology, MIT*. Massachusetts, USA: MIT.

Gassmann, O., Frankenberger, K. and Csik, M. (2012). The Business Model Navigator: 55 Models that Will Revolutionise Your Business (Pearson Education).

Gassmann, Oliver, Frankenberger, K. and Csik, M. (2014). The St. Gallen Business Model Navigator. Working Paper, University of St. Gallen.

Goldman, S. L., Nagel, R. N. and Preiss, K. (1995). Agile Competitors and Virtual Organisations (New York: Van Nostrand Reinhold).

Gordijn, J. (2002). Value-based requirements engineering – Exploring innovative E-commerce ideas, PhD thesis, Vrije Universiteit Amsterdam, The Netherlands.

Granovetter, M. S. (1973). "The Strength of Weak Ties," *The American Journal of Sociology*; 78, 6: 1360–1380.

Håkansson, H. (1980). "Marketing Strategies on Industrial Markets: A Framework Applied to a Steel Producer," *European Journal of Marketing*; 14, 5/6: 365–377.

Håkansson, H. (ed.) (1982). Internal Marketing and Purchasing of Industrial Goods – An Interaction Approach (New York: Wiley).

Håkansson, H. and Snehota, I. (1990). No Business is an Island: The Network Concept of Business Strategy (Uppsala: Uppsala University).

Håkansson, H. and Snehota, I. (1995). Developing Relationships in Business Network (Routledge).

Hamel, G. (2000). Leading the Revolution (Boston: Harvard Business School Press).

Hamel, Gary, (2001). "Leading the Revolution: An Interview with Gary Hamel," *Strategy & Leadership*; 29, 1: 4–10, https://doi.org/10.1108/10878570110367141.

Hamel, G. and Prahalad, C. K. (1994). "Competing for the Future," *Harvard Business Review* 72, 4: 122–128.

Hammer, M. (1990). "Business Process Reengineering Work; Don't Automate Obliterate," *Harvard Business Review*; July/August: 104–112.

Helg, R. (1999). "Italian Districts in the Industrial Economy," *Liucs Paper n. 68 Seria Economia e Empresa.*

Henry, W. R. and Haynes, W. W. (1978). Managerial Economics: Analysis and Cases (Dallas, Texas: Business Publications, Inc.).

Innovation Center Denmark (n.d.). Available at: http://icdk.um.dk/en/.

Johnson, M. W. (2010). Seizing the White Space. Business Model Innovation for Growth and Renewal (Boston, MA: Harvard Business Press).

Johnson, M. W., Christensen, C. M. and Kagermann, H. (2008). "Reinventing your Business Model. *Harvard Business Review*; 86: 50–59.

Kim, W. C. and Mauborgne, R. (2005). *Blue Ocean Strategy* (Harvard Business School Press).

Kirkeby, O. F. (2000). Management Philosophy – A Radical Normative Perspective Samfundslitteratur. (Heidelberg: Springer Verlag).

Kirkeby, O. F. (2003). Organisationsfilosofi Samfundslitteratur, København. English translation: Mette Morsing, Christina Thyssen (eds). New Forms of Loyalty. Corporate Values and Responsibility: the case of Denmark. Copenhagen: Samfundslitteratur, pp. 104–109.

Kotler, P. (1984). Principles of Marketing (Englewood Cliffs, NJ: Prentice Hall).

Krebs, C. J. (2009). Ecology: The Experimental Analysis of Distribution and Abundance. 6th ed. (San Francisco: Benjamin Cummings).

Kremar, H. (2011). "Innovation, Society and Business: Internet-based Business Models and their Implications," *Paper prepared for the 1st Berlin Symposium on Internet and Society, Oct. 25–27.*

Langager, C. (2010). The Industry Handbook. Available at: www.investopedia.com/features/industryhandbook/.

Li, M. et al. (2012). Neffics Delivery D 3.3. October 2012. Available at: www.neffics.eu.

Linder, J. and Cantrell, S. (2000). Changing Business Models: Surfing the Landscape (Cambridge, MA: Accenture Institute for Strategic Change).

Lindgreen, A. (2001). "A Framework for Studying Relationship Marketing Dyads," *Qualitative Market Research—An International Journal*; 4, 2: 75–87.

Lindgreen, A. and Wynstra, F. (2005). "Value in Business Markets: What do we Know? Where are we Going?" *Industrial Management*; 34: 732–748.

Lindgreen, A., Antioco, M. D. J. and Beverland, M. B. (2003). "Contemporary Marketing Practice: A Research Agenda and Preliminary Findings," *International Journal of Customer Relationship Management*; 6, 1: 51–72.

Lindgren, P. (2011). NEW Global ICT-based Business Models (River Publishers).

Lindgren, P. (2012). "Business Model Innovation Leadership: How do SMEs Strategically Lead Business Model Innovation?" *International Journal of Business Management*; 7, 14: 181–199.

Lindgren, P. (2016a). "Advance Business Model Innovation" *Journal of Personal Wireless Technology*. Springer Verlag.

Lindgren, P. (2016b). The Business Model Ecosystem (River Publishers).

Lindgren, P. (2017). Network Based High Speed Product Development (River Publishers). DOI: 10.13052/rp-9788793519053.

Lindgren, P. and Abdullah, A. M. (2013). "Conceptualizing Strategic Business Model Innovation Leadership for Business Survival and Business Model Innovation Excellence," *Journal of Multi Business Model Innovation and Technology*; 1, 3: 115–124.

Lindgren, P. and Bandsholm, J. (2016). "Business Model Innovation from a Business Model Ecosystem Perspective." *Journal of Multi Business Model Innovation and Technology*, 4, 2: 51–70.

Lindgren, P. and Clemmensen, S. (2008). "Preparing for a New Market Reality: Green Business Model Innovation in the Danish Household Sector," *Paper presented at 35th International Small Business Congress, Belfast, Ireland.*

Lindgren, P. and Dreisler, P. (2002). Produktudvikling i netværk: en kombination af forskellige "thought worlds" (AUBret).

Lindgren, P. and Jørgensen, R. (2012). "Towards a Multi Business Model Innovation Model," *Journal of Multi Business Model Innovation and Technology*; 1, 1: 1–22.

Lindgren, P. and Saghaug, K. M. (2012). Business Model and Intellectual Capital: How to Release Intellectual Capital from SME's BM's. IFKAD-KWCS, 2012.

Lindgren, P., Jørgensen, R., Saghaug, K. F. and Taran, Y. (2011). "Towards a Sixth Generation of Business Model Innovation Models," *CINet Conference. Aarhus, Denmark.* Continuous Innovation Network (CINet).

Lindgren, P. and Rasmussen, O. H. (2012). Business Model Innovation Leadership and Management. *Int. J. Bus. Manage.* 7: 14.

Lindgren, P. and Rasmussen, O. H. (2013). "The Business Model Cube," *J. Multi Bus. Model Innov.*; 135–182.

Lindgren, P., Rasmussen, O. H. and Saghaug, K. F. (2013). "Business Models Relations to Intellectual Capital – How to Release Intellectual Capital from Business BM's Relations?" in *Proceedings of the IFKAD Conference*, Zagreb.

Lindgren, P., Rasmussen, O. H. and Saghaug, K. F. (2014). ""Seeing" and "Sensing" Intellectual Capital in and between Different Business Model Eco Systems," in *Proceedings of the IFKAD Conference*, Matera.

Lindgren, P., Saghaug, K. F. and Clemmensen, S. (2009). "The Pitfalls of the Blue Ocean Strategy Canvas: The Importance of Value Related to the Strategy Canvas," in *Proceedings of the 10th International CINet Conference: Enhancing the Innovation Environment*. Vol. CD-rom Continuous Innovation Network (CINet), 588–601.

Lindgren, P., Saghaug, K. F. and Clemmensen, S. (2010). "The Pitfalls of the Blue Ocean Strategy: Implications of the Six Paths Framework," in *Proceedings CI Net 2010.*

Lindgren, P., Søndergaard, M. K., Nelson, M. and Fogg, B. J. (2013). "Persuasive Business Models," *Journal of Multi Business Model Innovation and Technology*; 1: 71–100.

Lindgren, P., Taran, Y. and Boer, H. (2010). "From single firm to network based business model innovation," *Int. J. Entrepr. Innov. Manage*; 12: 122–137.

Luthy, D. H. (1998). "Intellectual capital and its measurement," in *Proceedings of the Asian Pacific Interdisciplinary Research in Accounting Conference (APIRA)*, Japan. Available at: www.bus.osaka-cu.ac.jp/apira98/archives/htmls/25.htm.

Magretta, J. (2002). "Why Business Models Matter," *Harvard Business Review*; 80, 5: 86–92.

Malone, T. W., Weill, P., Lai, R. K., D'Urso, V. T., Herman, G., Apel, T. G. and Woerner, S. L. (2006). Do Some Business Models Perform Better than Others? MIT Working Paper 4615-06.

Markides, C. (2008). Game-changing Strategies: How to Create New Market Space in Established Industries by Breaking the Rules (San Francisco, CA: Jossey-Bass).

Markides, C. (2013). "Business Model Innovation: What can the Ambidexterity Literature Teach us? *Acad. Manag. Perspect*; 27: 313–323.

Markides, C. and Charitou, C. (2004). "Competing Dual Business Models: A Contingency Approach," *Academy of Management Executive*; 18, 3: 22–36.

MBIT (2014). MBIT Strategy Document 2014–2019. Available at: http://btech.au.dk/forskning/forskningsgrupper/mbit/.

Morris, M., Schmindelhutte, M. and Allen, J. (2003). "The Entrepreneur's Business Model: Toward a Unified Perspective," *Journal of Business Research*, 58, 6: 726–735.

Neffics (2012). Available at: www.Neffics.eu.

Neumann, J. v. (1928). “Zur Theorie der Gesellschaftsspiele,” *Mathematische Annalen*; 100, 1: 295–320, doi:10.1007/BF01448847; English translation: Tucker, A. W. and Luce, R. D., eds. (1959). “On the Theory of Games of Strategy,” *Contributions to the Theory of Games*; 4: 13–42.

Neumann, J. v. and Morgenstern, O. (1944). Theories of Games and Economic Behavior (Princeton University Press).

Odum, E. P. (1953). Fundamentals of Ecology (Philadelphia: W. B. Saunders).

OMG (2015). Available at: http://www.omg.org/spec/VDML/.

Osterwalder, A. (2011). Available at: https://www.youtube.com/watch?v=jMxHApgcmoU.

Osterwalder, A. (2014). Value Proposition Design (Strategyzer) – How to Create Products and Services Customers Want (Hoboken, NJ: Wiley & Sons).

Osterwalder, A. and Pigneur, Y. (2010). Business Model Generation (New Jersey: John Wiley & Sons, Inc.).

Osterwalder, A., Pigneur, Y. and Tucci, L. C. (2005). “Clarifying Business Models: Origins, Present, and Future of the Concept,” *Communications of AIS*; 16: 1–25.

Osterwalder, A. and Pigneur, Y. (2002). “An e-Business Model Ontology for Modeling e-Business,” *15th Bled Electronic Commerce Conference*, Bled, Slovenia.

Payne, A. and Holt, S. (1999). A Review of the Value Literature and Implications for Relationship Marketing. *Aust. Mark. J.* 7: 41–51.

Peng, M. W. (2010). Global (Mason, OH: South-Western).

Penrose, E. T. (1959). The Theory of the Growth of the Firm (New York: John Wiley).

Peters, Tom (1997). *The Circle of Innovation: You Can’t Shrink Your Way to Greatness*. (Hodder and Stoughton).

Peters, T. and Waterman, R. H. (1982). In Search of Excellence (New York: Harper & Row).

Peterovic, O., Kittl, C. and Teksten, R. D. (2001). Developing Business Models for eBusiness. International Conference on Electronic Commerce 2001, Vienna, October 31 – November 4.

Pleikys, S. (2012). Business Model Digitalization with the Case Study of Vlastuin. Master Thesis, International Technology Management, Aalborg University. Available at: http://projekter.aau.dk/projekter/files/63482301/BM_v05.31_fixed.pdf.

Porter, M. E. (1985). Competitive Advantage: Creating and Sustaining Superior Performance (Brighton, MA: Harvard Business Review).

Porter, M. E. (1996). “What is strategy?” *Harvard Business Review*; November–December: 61–78.

Porter, M. E. (1998). “Clusters and the New Economics of Competition,” *Harvard Business Review*, 76, 6: 77–90.

Porter, M. E. (2001). "Strategy and the Internet," *Harvard Business Review*; 3, 79: 62–79.

Porter, M. E. and Kramer, M. R. (2011). "Creating Shared Value," *Harvard Business Review*; 89, January–February 2011: 62–77.

Prahalad, C. K. and Hamel, G. (1990). "The Core Competence of the Corporation," *Harvard Business Review*; May–June, 79–91.

Provan, K. G. (1983). "The Federation as an Interorganizational Linkage Network," *The Academy of Management Review*; 8, 1: 79.

Provan, K. G., Fish, A. and Sydow, J. (2007). "Interorganizational Networks at the Network Level: A Review of the Empirical Literature on Whole Networks," *Journal of Management*; 33, 3: 479–516.

Provan, K. G. and Kenis, P. (2008). "Modes of Network Governance: Structure, Management, and Effectiveness," *Journal of Public Administration Research and Theory*; 18, 2: 229–252.

Rappa, M. (2001). Managing the digital enterprise - business models on the web, North Carolina State University. http://digitalenterprise.org/models/models.html [accessed June 2004].

Rasmussen, O. H. and Lindgren, P. (2015). "Understanding Relations, Relation Axiom, Persuasive Technologies and Sustainable Business Models," *International Conference on Wireless Communications, Vehicular Technology Information Theory, Aerospace & Electronic Systems.* IEEE Press.

Rasmussen, O. H. and Lindgren, P. (2016). "Two Black Boxes: Understanding the Coherence between Business Models & Business Model Eco Systems – A Contribution Toward a Definition of the Object for Business Model Innovation and the question of 'Where to Look'," *Journal of Multi Business Model Innovation and Technology*; 3, 3: 67–132.

Rasmussen, O., Lindgren, P. and Saghaug, K. (2014). Business Model Ecosystems and Interlectual Capital II - Why is Interlectual capital from business BM relations not released from a general Veblian framework condition perspective? Proceedings IFKAD 2014 Zagreb.

Ravn, J. (2012). Access2innovation: Netværksbaseret forretningsmodel innovation målrettet nødhjælps- og udviklingsbistanden. PhD Thesis, Institut for Planlægning, Aalborg Universitet.

Ravn, J., Slavensky, A., Taran, Y. and Lindgren, P. (2010). "The Conceptual Understanding of BoP Business Model(s)," *Paper presented at 7th ICCSR Symposium.* Corporate Social Innovation and Sustainable Community Development, Nottingham.

Ravn, J., Kroghstrup Nielsen, M. and Lindgren, P. (2009). "Network Based Business Model Innovation: Targeting the BOP Market," in *Proceedings of the 10th International CINet Conference: Enhancing the Innovation Environment.* Vol. CD-rom Continuous Innovation Network (CINet), 838–850.

Reichheld, F. F. (1993). "Loyalty-Based Management," *Harvard Business Review*; 71, 2: 64–73.

Rumelt, D. P. (1984). Towards a Strategic Theory of the Firm. Alternative Theories of the Firm. Elgar Reference Collection. International Library of Critical Writings in Economics, Vol. 154 (Cheltenham, UK and Northampton, MA: Elgar), 286–300.

Russell, M. (2011). Neffics workshop.

Russell, M. (2012). "Presentation Stanford University 2010 at the EU and US FinES Conference on Emerging Business Models," River Publisher and Martha Russell. Available at: http://www.youtube.com/watch?v=RrEigval78.

Saghaug, K. F. (2015). Revelation and Innovation of Value: Interacting between Revelatory Theology, Artistic Creativity and Small Business Owner's practice in a Business Model Innovation Context. PhD Dissertation, Center for Industrial Production, Aalborg University.

Saghaug, K. F. and Lindgren, P. (2010). "Implementing New Strategies of Operations in the Intersections of Academia and SMEs – With Special Focus on Human Beings as 'Differences' Acting on Relations Towards Meaningfulness," in *Proceedings of the 16th International Annual EurOMA Conference: Implementation – Realizing Operations Management Knowledge*, Gothenburg.

Sanchez, R. (1996). "Strategic Product Creation: Managing New Interactions of Technology, Markets and Organizations," *European Management Journal*; 14, 2: 121–138.

Sanchez, R. (2000). "Product, Process, and Knowledge Architectures in Organizational Competence," *Research Working Paper*, November. Oxford University Press.

Sanchez, R. (2001). Modularity, Strategic Flexibility, and Knowledge Management (Oxford: Oxford University Press).

Sandberg B. (2007). "Enthusiasm in the Development of Radical Innovations," *Creativity and Innovation Management*; 16, 3: 265–273.

SB (2009). International Auditing and Assurance Standards Board. "Audit Considerations In Respect of Going Concern In The Current Economic Environment," 1–11. Available at: http://www.ifac.org/system/files/downloads/IAASB_Staff_Audit_Practice_Alerts_2009_01.pdf.

Schulze, Ernst-Detlef; Erwin Beck; Klaus Müller-Hohenstein (2005). Plant Ecology (Berlin: Springer).

Selz, D. (1999). Value Webs: Emerging Forms of Fluid and Flexible Organizations. Thinking, Organizing, Communicating and Delivering Value on the Internet. Dissertation, University of St. Gallen, Switzerland, Bamberg: Difo – Druck OHG.

Shafer, S. M., Smith, H. J. and Linder, J. C. (2005). "The Power of Business Models," *Business Horizons*; 48, 3: 199–207.

Skarzynski, P. and Gibson, R. (2008). Innovation to the Core (Boston, MA: Harvard Business School Publishing).

Stewart, D. W. and Zhao, Q. (2000). "Internet Marketing, Business Models, and Public Policy," *Journal of Public Policy & Marketing*; 19: 287–296.

Sveiby, K. (2001). "Knowledge-based Theory of the Firm to Guide Strategy Formulation," *Journal of Intellectual Capital*; 2, 4. Available at: http://www.sveiby.com/articles/KnowledgeTheoryofFirm.pdf.

Sveiby, K. (2010). Methods for Measuring Intangibles. Available at: www.sveiby.com/articles/IntangibleMethods.htm.

Taran, Y. (2009). "Theory Building: Towards an Understanding of Business Model Innovation Processes," *Druid-Dime Academy 2009 PhD Conference*. Aalborg, Denmark: Centre for Industrial Production.

Taran, Y. (2011). Rethinking it All: Overcoming Obstacles to Business Model Innovation. PhD Thesis, Center for Industrial Production, Aalborg University.

Taran, Y., Boer, H. and Lindgren, P. (2015). "A Business Model Innovation Typology," *Decision Sciences*; 46: 301–331.

Teece, D. J. (2010). "Business Models, Business Strategy and Innovation," *Long Range Planning*; 43, 2–3: 172–194.

Teece D. J. (2012). "Dynamic Capabilities: Routines versus Entrepreneurial Action," *Journal of Management Studies*; 49, 8: 1395–1401.

Teece, D., Pisano, G. and Shuen, A. (1997). "Dynamic Capabilities and Strategic Management," *Strategic Management Journal*; 18, 7: 509–533.

Tillich, P. (1951). Reason and Revelation: Being and God (Chicago, IL: University of Chicago Press).

Tillich, P. (1990). Main Works/Hauptwerke bd 2. Writings in the Philosophy of Culture – Kulturphilosophische Schriften (New York, NY: Walter de Gruyter).

Timmers, P. (1998). "Business Models for Electronic Markets," *Journal on Electronic Markets*; 8, 2: 3–8.

Turban, E. (2003). Electronic Commerce: A Managerial Perspective (Pearson International).

Venkatraman, N. and Henderson, J. C. (1998). "Real Strategies for Virtual Organizing," *Sloan Management Review*, 40, 1: 33–48.

Vervest, P. et al. (2005). Smart Business Networks (Heidelberg-New York: Springer).

Von Hippel, E. (2005). *Democratizing Innovation* (Cambridge, MA: MIT Press).

Wadsworth, Y. (1998). What is Participatory Action Research? Available at: https://www.google.dk/search?source=hp&ei=RzZ7WrXSJ4SPmgXug5DoBQ&q=Wadsworth%2C+Y.+%281998%29&oq=Wadsworth%2C+Y.+%281998%29&gs_l=psy-ab.3..0i22i30k1.3558.3558.0.4064.3.2.0.0.0.0.106.106.0j1.2.0....0...1c.2.64.psy-ab..1.2.232.6..35i39k1.126.fHOPG5RgI2U.

Walter, A., Ritter, T. and Gemünden, H. G. (2001). "Value Creating in Buyer-seller Relationships," *Industrial Marketing Management*; 30, 4: 365–377.

Weill, P. and Vitale, M. R. (2001). Place to Space (Boston: Harvard Business School Press).

Wernerfelt, B. (1984). "The Resource-Based View of the Firm," *Strategic Management Journal*; 5, 2: 171–180. doi:10.1002/smj.4250050207.

Wernerfelt, B. (1995). "The Resource-Based View of the Firm: Ten Years After," *Strategic Management Journal*; 16, 3: 171–174. doi:10.1002/smj.4250160303.

Whinston, A. B., Stahl, D.O. and Choi, S. (1997). The Economics of Electronic Commerce (Indianapolis, IN: Macmillan Technical Publishing).

WIB (2013). Available at: http://interreg-oks.eu/webdav/files/gamla-projektbanken/se/Menu/Projektbank+2007-2013/Projektlista-Kattegat-Skagerrak/Women+in+Business.html.

William, H. (2011). "Going-Concern Assumption: Its Journey into GAAP," *The CPA Journal*; February: 26–28.

Williams, M. (2000). "Interpretivism and Generalisation," *Sociology*; 34, 2: 209–224.

Willis, A. J. (1997). "The Ecosystem: An Evolving Concept Viewed Historically," *Functional Ecology*; 11, 2: 268–271.

Woodruff, R. B. (1997). "Customer Value: The Next Source of Competitive Advantage," *Journal of Academy of Marketing Science*; 25, 2: 139–153.

Wouters, M. J. F., Anderson, J. C. and Wynstra, J. Y. F. (2005). The Adoption of Total Cost of Ownership for Sourcing Decisions (Amsterdam: Elsevier).

Zeithaml, V. A. (1988). "Consumer perceptions of price, quality, and value: a means-end model and synthesis of evidence," *Journal of Marketing*; 52, 3: 2–22.

Zott, C. and Amir, R. (2002). Measuring the Performance Implications of Business Model Design: Evidence from Emerging Growth Public Firms (Fontainebleau, France: INSEAD).

Zott, C., and Amit, R. (2009). "The business model as the engine of network-based strategies," in *The Network Challenge*, eds P. R. Kleindorfer and Y. J. Wind (Upper Saddle River, NJ: Wharton School Publishing), 259–275.

Zott, C., Amit, R. and Massa, L. (2010). The Business Model: Theoretical Roots, Recent Developments, and Future Research (Barcelona: IESE Business School University of Navarra).

Zott, C., Amit, R. and Massa, L. (2011). The Business Model: Recent Developments and Future Research. Available at: http://ssrn.com/abstract=1674384.

Index

L

M

N

O

P

Biographies

Professor Peter Lindgren
Aarhus University
Denmark

Peter Lindgren PhD holds a full Professorship in Multi Business Model and Technology Innovation at Aarhus University and has researched and worked with network-based high speed innovation since 2000. He was Head of Studies for the Master's Degree in Engineering – Business Development and Technology – at Aarhus University from 2014 to 2016. He is the author of several articles and books about business model innovation in networks and emerging business models. Peter has been a researcher at Politechnico di Milano in Italy (2002–2003), Stanford University, USA (2010–2011), University Tor Vergata, Italy and during 2007–2010 was the founder and Centre Manager of the International Center for Innovation (www.ici.aau.dk) at Aalborg University. He works today as a researcher in many different multi business model and technology innovation projects and knowledge networks, amongst others E100 (www.entovation.com/kleadmap), the Stanford University project Peace Innovation Lab (http://captology.stanford.edu/projects/peace-innovation.html), The Nordic Women in Business project (www.womenin business.dk), the Center for TeleInFrastruktur (CTIF) at Aalborg University (www.ctif.aau.dk), and the EU FP7 project about "multi business model innovation in the clouds" (www.Neffics.eu). He is co-author of several books. He has an entrepreneurial and interdisciplinary approach to research and has initiated several Danish and international research programmes. He is founder of the MBIT Lab and research group and is co-founder of CTIF Global Capsule (www.ctifglobalcapsule.com).

Peter's research interests are multi business model and technology innovation in interdisciplinary networks, multi business model typologies, sensing and persuasive business models.

Ole Horn Rasmussen is Postdoc at Aarhus University. His research interests are within the scientific field multi business model innovation, co-existence of old and new technologies, economics and the economic process. Before commitment to Aarhus University he was Post-Doctoral Fellow at the Department of Mechanical and Manufacturing Engineering, Aalborg University, Denmark. He has been Ph.D and Researcher at the Department of Economics, Politics and Public Administration and Department of Business and Economics at Aalborg University. His research interest ranges from (i) Business Model Innovation (ii) Strategic Business Model Innovation, Scenario Modeling, Business Accelerator and Sustainability (iii) Business Model Ecosystems (iv) Business Models, Engineering, Interdisciplinarity and Research Ontology, Epistemology and Methodology (v) Economics and the Economic Process. His empiric research methodology is a paticipative action research approach. The aim is to innovate new models and methods for integrated technology and business model innovation across product, service, production and process technology platforms in mind bothering multiple spin-outs. He works across industries, companies, public and private in order to create radically new solutions and integrable knowledge of those involved partners. To master the total integrated technology and business model innovation process from idea, concept, prototype implementation to operation, commercialization and bottom line is a prime goal. The results are incorporated into university educational programs, courses and training modules for students and Ph.D Schools.